Physics Calculations For GCSE & IGCSE

Brian Mills CPhys MInstP

Copyright © 2014 Brian Mills

All rights reserved.

ISBN: 978-0-9931501-0-4

DEDICATION

This book is dedicated to my lovely wife Joanne and my wonderful children Connor and Libby.

CONTENTS

Acknowledgments   i

	Introduction	Pg 3
1.	Powers of 10	Pg 4
2.	Prefixes	Pg 5
3.	Significant figures (s.f.)	Pg 6
4.	SI units	Pg 7
5.	Changing the subject of the formula	Pg 8
6.	Advice when solving Physics formula	Pg 13
7.	Speed	Pg 14
8.	Acceleration	Pg 18
9.	Weight	Pg 22
10.	F = ma	Pg 25
11.	Momentum	Pg 30
12.	Force = rate of change of momentum	Pg 33
13.	Conservation of momentum	Pg 37
14.	Work done	Pg 42
15.	Kinetic energy	Pg 46
16.	Gravitational potential energy	Pg 50
17.	Conservation of energy	Pg 54
18.	Moments	Pg 56
19.	Law of moments	Pg 59

20.	$F = k \times e$	Pg 64
21.	Centripetal force	Pg 67
22.	Density	Pg 71
23.	Pressure	Pg 74
24.	Pressure variation with depth in fluids	Pg 77
25.	Hydraulics	Pg 80
26.	Boyle's Law	Pg 83
27.	Pressure Law	Pg 86
28.	Specific heat capacity	Pg 90
29.	Latent heat	Pg 94
30.	$V = I \times R$	Pg 99
31.	Series circuits	Pg 101
32.	Parallel circuits	Pg 107
33.	$Q = I \times t$	Pg 111
34.	$E = V \times Q$	Pg 113
35.	$P = V \times I$	Pg 116
36.	$E = P \times t$	Pg 119
37.	Paying for electricity	Pg 122
38.	$E = V \times I \times t$	Pg 124
39.	Efficiency	Pg 126
40.	Transformers	Pg 129
41.	$v = f \times \lambda$	Pg 133
42.	$T = 1/f$	Pg 136
43.	Refractive Index	Pg 138
44.	Total internal reflection and critical angle	Pg 143

45.	$P = 1/f$	Pg 146
46.	$1/u + 1/v = 1/f$	Pg 150
47.	Magnification	Pg 155
48.	Half-life	Pg 157
49.	Graphs	Pg 161
50.	Velocity	Pg 175
51.	Answers to questions	Pg 178

ACKNOWLEDGMENTS

I would like to thank all of the people that have taught me Physics and given me inspiration to study and now teach such a fascinating subject. I would also like to thank all of my students, past, present and future.

Introduction

This book has been written to help students perform calculations successfully at GCSE & IGCSE level.

When using the book it is recommended that you first read sections 1-6 at the start of the book, which explains powers of 10, prefixes, significant figures, SI units, changing the subject of a formula (rearranging formulae) and tips for answering Physics calculation problems. You can then go to whatever Physics formula you want to attempt.

Read carefully the introduction about each formula. This will include the units that each quantity in the formula is measured in, how to change the subject of the formula and may also include tips on how to avoid common mistakes that students make. Once you have done this, you can check your understanding by working through the questions for each respective formula.

The book explains the formulae in detail used by the main examination boards at GCSE & IGCSE. Each section explains the formula and how to rearrange it, with technical terms and the Physics explained. There are also plenty of worked examples and questions at the end of each section to test your knowledge. You will need to check your exam board specification for the formulae that you will be examined on, although you may also want to work through all of them to improve your Physics.

Answers to the questions are at the back of the book so that you can check your progress.

Good luck with all of the Physics problems and I hope that you get a lot of satisfaction from answering them correctly.

1. Powers of 10

There are times when you will be given quantities in a question as a power of 10. For example the speed of light in a vacuum is given as 3×10^8 m/s (3 times 10 to the power of 8 metres per second). It is important that you understand what this means.

If you get a single digit number positive number (1 to 9) times by 10 to the power of something, all you need to do is write down the number, and whatever the power is, in this case 8, you write down that many zero's. So, in the case of 3×10^8 m/s, you write down the number 3 and because the power is 8, after the 3 you write down 8 zeros. Therefore, 3×10^8 m/s is the same as 300,000,000 m/s.

Giving another example, imagine that an electron has a speed of 4×10^6 m/s. Following our method, the number will be the same as 4,000,000 m/s. (write down the number and whatever the power is, in this case 6, you write down that many zero's after the single digit number).

Alternatively, a method which works for any type of number is the decimal point method:
Taking 3×10^8 m/s as an example, write the number 3 as 3.0 because we need a decimal point. The power of 10 is 8. If the power is a positive number which it is in this instance, you move the decimal point 8 (in this example) places to the **right**. Whatever the power of 10 is, the decimal point moves that many places. Therefore 3×10^8 m/s becomes 300,000,000 m/s. The decimal point has moved 8 places to the **right!**
Let's now consider an electron which has a speed of 4×10^6 m/s. Write down the number 4 as 4.0 and move the decimal point to the **right** the same number of times as the power (in this case 6). Therefore 4×10^6 m/s becomes 4,000,000 m/s.
Considering another example, if we have a force of 6.24×10^5 N, then the decimal point needs to be moved 5 places to the **right**. Our number then becomes 624,000N.

If you are given a power which is a negative number, you need to use the decimal point method in order to work it out. For example, imagine that we have a voltage of 20×10^{-3} V (20 times 10 to the power of minus 3 volts). You need a decimal point, so write down the number 20 as 20.0. Then because the power is a negative number, -3, you move the decimal point 3 places to the **left**. This will give you 0.020V.

Let's take a look at another example. The charge on the electron is 1.6×10^{-19} C (1.6 times 10 to the power of minus 19 coulombs). Coulombs is the unit of electrical charge. This time we have a decimal point already, so write down the number 1.6. Then move the decimal point 19 times to the **left**, because the power is -19. This can then be written as 0.00000000000000000016 C. As you can see, this is a very small number and is one of the reasons why we use powers of 10. It is a little tedious writing down all of those zero's. It is far easier to write the number as a power of 10.

In summary, if the power is a positive number the decimal point moves that many places to the right. If the power is a negative number, the decimal point moves that many places to the left.

I advise trying all of the above examples for yourself on a piece of paper to make sure that you agree and to also check your understanding. You can then have a go at the question below and check your answers with the ones at the back of the book.

Practice question

1) Write down the equivalent numbers for the following:
a) 4×10^7
b) 3×10^4
c) 5.37×10^6
d) 45×10^{-6}
e) 6.67×10^{-11}
f) 2.5×10^4
g) 5.1×10^{-6}
h) 7.14×10^8
i) 9.04×10^{-3}
j) 2×10^5

As your have just seen it is easier to write very large or very small numbers as a power of 10. As you will see in the next section, it is also useful to use prefixes.

2. Prefixes

Prefixes are used in Physics to make it easier to write very large or very small numbers. The table below shows the standard prefixes. The multiplier column gives 2 possible multipliers. You can either use the power of 10 given or the number in brackets. Both methods will give the same answer but in a different form. See examples 1, 2 & 3 overleaf.

Prefix	Abbreviation	multiplier
peta	P	$\times 10^{15}$ (x1,000,000,000,000,000)
tera	T	$\times 10^{12}$ (x 1,000,000,000,000)
giga	G	$\times 10^9$ (x 1,000,000,000)
mega	M	$\times 10^6$ (x 1,000,000)
kilo	k	$\times 10^3$ (x 1000)
centi	c	$\times 10^{-2}$ (x 0.01)
milli	m	$\times 10^{-3}$ (x 0.001)
micro	µ	$\times 10^{-6}$ (x 0.000001)
nano	n	$\times 10^{-9}$ (0.000000001)
pico	p	$\times 10^{-12}$ (0.000000000001)
femto	f	$\times 10^{-15}$ (x0.000000000000001)

Usually at GCSE & IGCSE you should only need the prefixes from giga to nano. Below are a few examples to explain prefixes.

Example 1

Consider a force of 50MN. This means that the force is 50 meganewtons. There are 2 easy ways to work out what the number is in newtons, N:
1) Replace the prefix (M) with the multiplier as a power of 10 from the table. 50MN becomes 50×10^6 N.
2) Replace the prefix (M) with the multiplier from the table. 50MN becomes $50 \times 1,000,000 = 50,000,000$N.

You can use whichever method you find easiest, although ideally it is better to be comfortable with both.

Example 2

Consider a current of 20mA. This means that the current is 20milliamps. In order to convert this to amps, A:
1) Replace the prefix (m) with the multiplier as a power of 10 from the table. 20mA becomes 20×10^{-3} A.
2) Replace the prefix with the multiplier from the table. 20mA becomes $20 \times 0.001 = 0.020$A.

Example 3

Consider a time of 30μs. This means a time of 30 microseconds. In order to convert this to seconds, s:
1) Replace the prefix (μ) with the multiplier as a power of 10 from the table. 30μs becomes 30×10^{-6} s.
2) Replace the prefix with the multiplier from the table. 30μs becomes $30 \times 0.000001 = 0.000030$ s.

Questions

1) Write 75kN in newtons, N.
2) Write 50μs in seconds, s.
3) Write 5MV in volts, V.
4) Write 4mA in amps, A.
5) Write 10GJ in joules, J.
6) Write 8mm in metres, m.
7) Write 6nC in coulombs, C.
8) Write 60cm in metres, m.

3. Significant figures (s.f.)

The significant figures in a number are all of the digits except any zero's before the first non-zero digit. Below are some examples, the significant figures for each number are in bold.

36, **3.6**, 0.0**36**, 0.00**36** and **3.6**x10^5 all have 2 significant figures.

5.05, **505**, 0.0**505** and **5.05**x10^4 all have 3 significant figures.

The number of significant figures indicates the precision with which the measurement was made. For example a measurement of 4.20mm is more precise than a measurement of 4.2mm. 4.20mm has been measured to the nearest 0.01mm while 4.2mm has been measured to the nearest 0.1mm.

When answering calculations, the answer needs to be given to the same precision as the least precise data in the question.
Consider the example below:
An athlete runs 400m in 44.56s. Calculate the speed.

$$speed = \frac{distance}{time}$$

$$s = \frac{d}{t}$$

$$s = \frac{400}{44.56}$$

Your calculator gives the answer

s = 8.976660682 m/s (metres per second)

This amount of precision is far greater than the precision of the data given in the question. The least amount of significant figures given in the question is 3 (400m), therefore the answer must be given to 3 significant figures.

This will give the answer to be s = 8.98 m/s.

Don't forget that if you look at the calculator answer
s = 8.976660682 m/s. The number after the 7 is a 6. The rule is round up if the number is 5 or above and round down if below 5. You would therefore be wrong if you wrote the answer to be 8.97 m/s.
The correct answer would be 8.98 m/s.

4. SI units

The formulae in this book will all use SI units (The system of international units). All quantities in a formula are to be measured in SI units unless told otherwise. They will be given after each formula. An example is given below.

Resultant force = mass x acceleration

(This formula can be abbreviated to: F = m x a or F = ma)

Resultant force is measured in newtons, N.
Mass is measured in kilograms, kg.
Acceleration is measured in metres per second squared, m/s^2.

Therefore the SI unit of:
Force is the newton, N.
Mass is the kilogram, kg.
Acceleration is the m/s^2 (metres per second squared).

All numbers must be put into each formula in the same units as stated. For example, if in a question you are given a force of 6kN for the above formula, the number that you need to put into the formula is 6000N or 6×10^3N otherwise you are going to get the answer wrong. The force must be in newtons, N, not kilonewtons, kN. This also highlights the importance of understanding prefixes and powers of 10!

5. Changing the subject of a formula

It is important to understand the basic rules of algebra in order to change the subject of a formula (rearrange a formula) successfully. Some of this will be dealt with now, while other bits will be dealt with while looking at specific formulae later on.

Consider the equation $V = I \times R$ (I will look at this formula in detail in section 30, including what the letters stand for. At the moment though I will just concentrate on how to rearrange it).

Let's firstly rearrange this formula to make **I** the subject of the formula:

$$V = I \times R$$

We need to get the **I** by itself. To do this we need to get rid of the **R** on the right hand side. To do this you have to do the 'inverse' or 'opposite' operation. What I mean by this is that the '**I** and **R**' are multiplied together, therefore, to get rid of the **R** from the right hand side of the formula you need to divide both sides by **R**. It is important to do it to **both sides** otherwise you are changing the formula. This will give:

$$\frac{V}{R} = \frac{I \times R}{R}$$

Looking at the right hand side of the equation the **R's** cancel (I will explain this after the next paragraph).
This will give:

$$\frac{V}{R} = I$$

The whole point of dividing both sides of the formula by **R**, was to get rid of the **R** from the right hand side of the formula. This is what we have achieved. It does not matter now which way you write this (because they are both equal to each other) and you would usually write:

$$I = \frac{V}{R}$$

Let's go back to explain why the **R's** cancelled.

$$\frac{V}{R} = \frac{I \times R}{R}$$

Looking at the right hand side of the formula only, let's make up some numbers for **I** and **R**. Let **I** = 5 and **R** = 10. This will give:

$$\frac{5 \times 10}{10}$$

There are 2 ways to look at this to prove that you are just left with the 5 (which is I):

1) 5 x 10 = 50. Then 50 ÷ 10 = 5
2) Both of the 10's will cancel out leaving just the 5.

The same rules apply to letters, so don't be confused by this. Therefore on the right hand side you are just left with **I**.

Note:
It is worth noting that:

$$I = \frac{V}{R}$$

Can be written also in the following ways:

$$I = V/R$$

Or,

$$I = V \div R$$

Let's now go back to the original formula, **V = I x R**, and make **R** the subject of the formula.

$$V = I \times R$$

Using the same method, this time you want to get **R** by itself. To do this we need to get rid of the **I** on the right hand side. Again, this will mean doing the 'inverse' or 'opposite' operation. What I mean by this is that the '**I** and **R**' are multiplied together, therefore to get rid of the **I** from the right hand side of the formula you need to divide both sides by **I**. Again, it is important to do it to **both sides** otherwise you are changing the formula. This will give:

$$\frac{V}{I} = R$$

I have missed out the middle step this time as you should now be okay with what I have done. If not, go back to the start of this section and re-read it. Hopefully you will then be fine.
Finally this can be written as:

$$R = \frac{V}{I}$$

Let's now consider another formula. I will look at the formula in detail in section 8, including what the letters stand for. At the moment though I will just concentrate on how to rearrange it.

$$a = \frac{v - u}{t}$$

Let's firstly rearrange this formula to make **t** the subject of the formula. We need to get the **t** above the line (in the position of the numerator), not below it (denominator). Due to the fact that '**v-u**' is divided by **t**, to get the **t** above the line you need to do the 'inverse' or 'opposite' operation. This means that you multiply both sides by **t**. This gives:

$$a \times t = \frac{(v - u)t}{t}$$

The **t**'s on the right hand side of the formula cancel to give:

$$a \times t = v - u$$

Note:
(Can you see that the **t** is now effectively in the position of the numerator (the top part of a fraction) even though we have not got a fraction. Before it was in the position of the denominator (the bottom part of a fraction). What I am trying to say is that if we look at the formula:

$$a = \frac{v - u}{t}$$

I would effectively be classing **a**, **v** and **u** to be in numerator positions and **t** in a denominator position, even though strictly speaking this would not be entirely correct as you need a fraction to have a numerator and a denominator. It is just helping me to explain which ones are above the line (**a**, **v** and **u**) and which ones are below it, **t**).

Referring back then to where we were:

$$a \times t = v - u$$

Now we need to get the **t** by itself and therefore get rid of the **a**. Due to **a** and **t** being multiplied together, in order to get rid of the **a** you must do the 'inverse' or 'opposite' operation, which in this case will be to divide both sides by **a**. This will give:

$$\frac{a \times t}{a} = \frac{v - u}{a}$$

The **a**'s on the left hand side of the formula cancel to give:

$$t = \frac{v - u}{a}$$

Now let's make **v** the subject of the formula. The first steps are similar to before, so I will go through them a little quicker.

$$a = \frac{v - u}{t}$$

Multiplying both sides by **t** will again give:

$$a \times t = v - u$$

In Mathematics, when 2 quantities are multiplied together you can also write it like this:

$$at = v - u$$

There is no need (if you don't want to), to put in the multiplication sign. 'at' means **a** multiplied by **t**.

Now if you look at the right hand side of the formula, we have '$v - u$'. We want to get the **v** by itself so we need to get rid of the **−u**. Again, you need to do the 'inverse' or 'opposite' operation. This means that you need to add **u** to both sides. This gives us:

$$at + u = v$$

This is then the same as:

$$v = at + u$$

Which is also the same as:

$$v = u + at$$

(Don't forget, it doesn't matter what way we add numbers together. For example, 5 + 4 = 9, 4 + 5 = 9, 9 = 5 + 4. The same applies to letters).

We have therefore made **v** the subject of the formula.
Let's look in detail at what we have done here. Adding **u** to both sides would have given us:

$$at + u = v - u + u$$

Looking at the right hand side of the formula we have $v - u + u$
The **−u** and **+u** cancel each other out. Imagine giving **u** the value of 5. Looking at $-u + u$, this would give us -5 +5 which is zero. On the right hand side you would then just be left with the **v**.
This then gave us:

$$at + u = v$$

Looking at the left hand side of the formula only we have '$at + u$'.
Imagine that 'at' gives us an answer of 10 if **a**=5 and **t**=2. If we then say that **u**=4. This means that $at + u$ will equal 10 + 4. In Mathematics it does not matter if you write this as 10 + 4, or 4 +10. You will still get the same answer. So, finally you can write:

$$at + u = v$$

As

$$v = u + at$$

Before we make **u** the subject of the formula there is something that I need to explain. Firstly I need to explain BIDMAS.
BIDMAS is used in mathematics for the order of operations. Mathematical operations need to be done in the following order:
1st – **B**rackets
2nd – **I**ndices (This is just another word for powers, for example 10^2, $5^{1/3}$, $3^{1/2}$).
3rd – **D**ivision
3rd - **M**ultiplication
4th – **A**ddition
4th – **S**ubtraction

'Division and multiplication' have third equal priority and 'addition and subtraction' have fourth equal priority. This means that any 'division or multiplication' needs to be calculated before any 'addition or subtraction'.

Let's look at the formula and put some numbers in to help clear the situation. Let:
u = 5 m/s, **a** = 2 m/s² and **t** = 10s.

$$v = u + at$$

Which is the same as:

$$v = u + a \times t$$

The multiplication must be done first because it has a higher order of priority than addition. Therefore '$a \times t$' will equal 2 x 10 = 20. Then 5+20 = 25m/s. This is the correct answer.

A common mistake is to do '$u + a$' first which would be 5 + 2 = 7. Then multiply this by **t**. So, 7 x 10 = 70! **This is wrong!** The operations have been done in the wrong order. The addition has been done before the multiplication!

If we now go back to rearranging the formula:

$$v = u + at$$

and make **u** the subject of the formula. On the right hand side we have '$u + at$'. Here 'at' is added to the 'u'. 'at' would need to be multiplied together first before adding to the 'u'. Therefore in order to get the 'u' by itself, you must treat the 'at' as 'one' and so you must subtract 'at' from both sides of the formula. This gives:

$$v - at = u$$

Which is the same as:

$$u = v - at$$

If we check that we have rearranged this correctly let's substitute **v**, **a** and **t** into the formula from the numbers above.
v = 25 m/s (this was calculated), **a** = 2 m/s² and **t** = 10s.

We then have:
$$u = 25 - 2 \times 10$$

$$u = 25 - 20$$

(remember the multiplication has to be done first in this case)

$$u = 5 m/s$$

and this agrees with the value that we gave **u** earlier, so we have rearranged the formula correctly.

In summary, the formula:
$$a = \frac{v - u}{t}$$

can be rearranged as shown below to make **t**, **v** and **u** the subject of the formula respectively:

$$t = \frac{v - u}{a}$$

$$v = u + at$$
And
$$u = v - at$$

It is very important that you are able to rearrange formulae and I would recommend reading this section again to check your understanding. That's it for now on changing the subject of formulae. More will be done when we look at individual Physics formulae later on. Below are some questions that you can attempt to check your understanding of this section. Answers are at the back of the book. Good luck, and again, please re-read this section if you need to!

Questions

Rearrange the following formulae.

a) Make **a** the subject of the formula $F = ma$.
b) Make **t** the subject of the formula $Q = It$.
c) Make **g** the subject of the formula $W = mg$.
d) Make **t** the subject of the formula $s = \frac{d}{t}$
e) Make **d** the subject of the formula $s = \frac{d}{t}$

And now a few of a more advanced level

f) Make **s** the subject of the formula **v² = u² + 2as**
g) Make **t** the subject of the formula $s = \frac{1}{2}(u + v)t$

6. Advice when solving Physics problems

When solving problems I recommend following the 5 steps given below:

1) Make a list of the quantities given in the question.
2) Write down the required formula.
3) If needed, rearrange the formula.
4) Substitute the numbers into the formula.
5) Write down the answer together with the correct units.

Consider the following question as an example in using these 5 steps:

A toy rocket of mass 2kg has a resultant force acting on it of 10N. Calculate the acceleration.

We will use the formula:

Resultant force = mass x acceleration (this formula will be looked at in detail in section 10)

This formula can be abbreviated to: **F = m x a** (or **F = ma**)
Where **F** is the resultant force measured in newtons, N.
m is the mass, measured in kilograms, kg.
a is the acceleration, measured in metres per second squared, m/s².

Step 1 - Make a list of the quantities given in the question.

m= 2kg
F = 10N
a = ?

Step 2 - Write down the formula required.

$$F = m \times a$$

Step 3 - If needed, rearrange the formula.

To make **a** the subject of the formula, we need to divide both sides by **m**. This will give:

$$\frac{F}{m} = a$$

Therefore,

$$a = \frac{F}{m}$$

Step 4 - Substitute the numbers into the formula.

$$a = \frac{10}{2}$$

Step 5 - Write down the answer together with the correct units.

$$a = 5 m/s^2$$

I recommend following this method every time you solve a problem, even if the problem seems really easy. It will mean you are less likely to make a mistake and also, if by chance you do make a mistake, because you have shown full working, you will pick up method marks in an exam question. It is far better if you do make a mistake to pick up maybe 2 marks out of say 3, rather than 0 out of 3. You need to show the examiner what you know and you can only do this on the exam paper. The examiner can't mark what is in your head or what is done on your calculator. **Show full working all of the time!**

Sections 1-6 have been written to help you with the rest of the sections in the book, which, section at a time will look at an individual Physics formula in detail. I recommend that when needed you dip back into these sections to refresh your memory on the topics that have been covered.

7. Speed

The formula for speed is given below:

$$speed = \frac{distance}{time}$$

Speed is measured in metres per second, m/s.
(In some text books you will also see this written as ms^{-1}. It doesn't matter which one you use, they both mean the same thing).

Distance is measured in metres, m.
Time is measured in seconds, s.

This formula can be written as:
$$s = \frac{d}{t}$$

You must remember when using the formula that all quantities must be expressed in the SI units that are given above!

First of all let's look at rearranging the formula.

$$s = \frac{d}{t}$$

If you want to make **d** the subject of the formula (or rearranging for **d**), you need to multiply **both** sides by **t**.
This will give:

$$s \times t = d$$

Therefore,

$$d = s \times t$$

If you want to make **t** the subject of the formula (or rearranging for **t**), using the above formula, $d = s \times t$, you now need to divide both sides by **s**. This will give:

$$\frac{d}{s} = t$$

Therefore,

$$t = \frac{d}{s}$$

You can now use the above formulae to solve problems on speed. I will give 3 examples before you have a go at the questions.

Example 1

An athlete runs 200m in 20s. Calculate the speed.

Listing the quantities from the question we have:

d = 200m
t = 20s
s = ?

$$s = \frac{d}{t}$$

$$s = \frac{200}{20}$$

s = 10 m/s

Example 2

A cyclist travels at 12 m/s for 1 minute. How far does he travel?

Firstly, all units need to be the same as those given at the start of this section. Time is measured in seconds in the formula, so we need to convert 1 minute into seconds. 1 minute = 60 seconds.

s = 12 m/s
t = 1 minute = 60 seconds
d = ?

$$s = \frac{d}{t}$$

$$d = s \times t$$

$$d = 12 \times 60$$

$$d = 720m.$$

Example 3

A car covers 10,000m while travelling at a constant speed of 80.0 km/h (kilometres per hour). How long does this take?

Firstly, all units need to be the same as those given for the speed formula at the start of the section.
Therefore you need to convert 80.0 km/h to m/s.

We will do this in stages. First we will convert it to m/h (metres per hour), then we will convert this to m/s.

Let's look at the speed of 80.0 km/h. The k stands for kilo, so we have to replace the k with either **x 1000** or **x 10³**. This means that 80.0 km/h becomes 80,000 m/h. We now need to convert m/h to m/s. There are 60s in 1 minute and 60 minutes in hour. In 1 hour this means that there will be 60 x 60 = 3600s in 1 hour.

$$\frac{80,000}{3600} = 22.2$$

So, 80.0km/h = 22.2 m/s.

If we now continue to solve the problem.

s = 22.2 m/s
d = 10,000m
t = ?

$$s = \frac{d}{t}$$

$$t = \frac{d}{s}$$

$$t = \frac{10,000}{22.2}$$

$$t = 450s$$

(Note that the least number of significant figures given in the question was 3, so the answer was given to 3 significant figures).

Sometimes with this formula it is possible that SI units will not be used. If this is the case, it is important that you are consistent with the units. For example, if speed is measured in kilometres per hour, **km/h**, then the distance must be measured in kilometres, **km** and the time must be measured in hours, **h**.

Answers to questions should always be given in SI units, unless told otherwise!

Example 4

A car travels at 50km/h for 2.5 hours. Calculate the distance travelled in kilometres.

s = 50km/h
t = 2.5h
d = ?

$$s = \frac{d}{t}$$

$$d = s \times t$$

$$d = 50 \times 2.5$$

$$d = 125 \text{km}$$

Note:
Velocity is speed in a given direction. If the velocity formula (which involves velocity, displacement and time) is on your specification this be dealt with in detail in section 51.

Now it is time for you to attempt some questions.

Questions.

1) An athlete runs 400m in 45.6s. Calculate the speed.
2) A Lorry covers 200km in 4 hours. Calculate the speed in km/h.
3) A woman walks 2.00km in 1200s. Calculate the speed.
4) A Cheetah runs at 25 m/s for 40s. How far does it travel?
5) A cyclist travels at 12.0 m/s for 180s. How far does he travel.
6) A cricket ball travels at 20 m/s for 50s. How far does it travel?
7) A greyhound covers 200m while travelling at 15.0 m/s. How long does this take?
8) Lava is flowing from a volcano at 2.30 m/s. If it flows 300m, how long does this take?
9) A lorry is travelling at 20 m/s for 30s. How far does it travel?
10) A snail crawls 12.0cm in 60.0s. Calculate the speed.
11) A formula 1 car travels at 375km/h for 40.0s. How far does it travel?
12) A car travels 160.0km while travelling at 80.0 km/h. How long does this take in hours ?
13) The distance from the Sun to the earth is 1.5 x 10^{11} m. How long will it take light to reach the earth from the Sun if the speed of light is 3.0 x 10^8 m/s?

14) A car covers 6.0 x 10^{5}m while travelling at 80.0 km/h. How long does the journey take?
15) The space shuttle is travelling at 17,500mph (miles per hour). How far will it travel in 3600s (1 hour)? (hint: 1mile is approximately 1609m).

8. Acceleration

The formula for acceleration is given by:

$$acceleration = \frac{change\ in\ velocity}{time}$$

Note:
Velocity is the speed in a given direction. For example:
Speed = 10 m/s
Speed = 30 m/s
Speed = 50 m/s
Velocity = 10 m/s north
Velocity = 36 m/s down
Velocity = 65 m/s east

As mentioned, velocity is speed in a given direction. If the velocity formula (which involves velocity, displacement and time) is on your specification this be dealt with in detail in section 51.
You may also find that when velocities are given in questions that the direction is not given. Do not worry about this in this section, again, see section 51.

Change in velocity can be calculated as follows:

Change in velocity = Final velocity – initial (start) velocity

The acceleration formula can therefore be written as:

$$a = \frac{v - u}{t}$$

a = acceleration (measured in metres per second squared, m/s^2).
(In some text books you will also see this written as ms^{-2}. It doesn't matter which one you use as they both mean the same thing).

v = final velocity (measured in metres per second, m/s).

u = initial (start) velocity (measured in metres per second, m/s).

t = time (measured in seconds,s).

Due to the fact that I rearranged this formula in section 5 (I suggest that you go back and check how to do this) I will now just write down what the formula rearranges to:

$$t = \frac{v - u}{a}$$

$$v = u + at$$

$$u = v - at$$

Note:
These formulae only apply to constant (steady) accelerations.

You can now use the above formulae to solve problems on acceleration. I will give 5 examples before you have a go at the questions.

Example 1

A ball accelerates from rest to 10 m/s in 0.50s. Calculate the acceleration.

It is important to realise that 'rest' means that the 'initial' or 'starting' velocity is 0 m/s!!

u = 0 m/s
v = 10 m/s
t = 0.50s
a = ?

$$a = \frac{v - u}{t}$$

$$a = \frac{10 - 0}{0.50}$$

$$a = \frac{10}{0.50}$$

a = 20 m/s²

Let's take a look at what we mean by an acceleration of 20 m/s². An easy way to look at it is rather than write the unit as m/s², write it as m/s/s (metres per second per second). Both of these units are identical. If we look then at an **acceleration of 20m/s/s**, what this means is that the **velocity changes by 20m/s every second!**

See the table below to see how the velocity would change with time if the ball continued at this constant acceleration.

t/s	0	1	2	3	4
v/ m/s	0	20	40	60	80

Example 2

A car accelerates at 5 m/s² from 4.0 m/s to 24.0 m/s. How long does this take?

a = 5.0 m/s²
u = 4.0 m/s
v = 24.0 m/s
t = ?

$$a = \frac{v - u}{t}$$

$$t = \frac{v - u}{a}$$

$$t = \frac{24.0 - 4.0}{5.0}$$

$$t = \frac{20.0}{5.0}$$

$$t = 4.0s$$

Hint:
When making the list of quantities as above, it is important to learn the units for each formula. A common mistake is to write down the acceleration as a velocity, for example, a = 24m/s. **This is wrong!** The units m/s, are a measure of velocity, not acceleration. Don't forget, acceleration is measured in m/s² and velocity is measured in m/s!

Example 3

A cyclist travelling at 6 m/s accelerates at 2 m/s² for 4s. Calculate the final velocity of the cyclist.

a = 2 m/s²
u = 6 m/s
t = 4s
v = ?

$$a = \frac{v - u}{t}$$

$$v = u + at$$

$$v = 6 + 2 \times 4$$

(remember **BIDMAS**, multiplication must be done before addition).

$$v = 6 + 8$$

$$v = 14 \text{ m/s}$$

Example 4

A cyclist accelerates at 3 m/s² for 4s until he reaches a velocity of 16 m/s. What was the initial velocity?

a = 3 m/s²
v = 16 m/s
t = 4s
u = ?

$$a = \frac{v - u}{t}$$

$$u = v - at$$

$$u = 16 - 3 \times 4$$

(remember **BIDMAS**, multiplication must be done before addition).

$$u = 16 - 12$$

u = 4 m/s

Example 5

A car is travelling at 30 m/s and comes to rest in 6s. Calculate the acceleration.

u = 30 m/s
v = 0 m/s
t = 6s
a = ?

$$a = \frac{v - u}{t}$$

$$a = \frac{0 - 30}{6}$$

$$a = -\frac{30}{6}$$

a = - 5 m/s²

The negative sign just indicates that we have a deceleration (slowing down) rather than an acceleration (speeding up).
Another way of expressing an acceleration of - 5 m/s², is to say that the **deceleration = 5 m/s²**. Therefore, in this example the velocity decreases by 5 m/s every second.

Questions

1) A cyclist accelerates from rest at 2 m/s² for 6s. Calculate the final velocity.
2) A car accelerates from 10 m/s to 30 m/s in 5s. Calculate the acceleration.
3) A lorry accelerates at 4 m/s² from rest to 32 m/s. How long does this take?
4) A train is accelerates at 2 m/s² for 15s until it reaches a velocity of 40 m/s. What was the initial velocity?
5) A motorbike travelling at 60 m/s comes to rest after decelerating at 10 m/s². How long does this take? **Hint:** You need to write down the acceleration as -10 m/s² in the formula.
6) A car has a constant acceleration of 3 m/s². If it starts from rest, what will be the velocity after 15s?
7) A Ferrari 430 Scuderia goes from 0 – 60.0 mph in 3.60s. Calculate the acceleration assuming it was constant.
8) In a 100m race an athlete accelerates from rest to 10 m/s in 4.0s. Calculate the acceleration.
9) A Porche 911 is travelling at 43.4 m /s when the driver sees a 30 mph (13.4 m/s) sign. If the acceleration is -10 m/s², how long will it take for the driver to come down to the speed limit?
10) The Saturn V rocket used in the Apollo space missions reached a speed of 2.756 x 10³ m/s, 161s after lift-off. Calculate the acceleration, assuming it to be constant.

9. Weight

Before we take a look at the weight formula, it is important that you understand the difference between mass and weight.

Weight is the force of gravity acting on an object. It is a force and it is measured in newtons, N.

Mass is the amount of matter contained within an object. It is measured in kilograms, kg.

In everyday life the words mass and weight are used interchangeably for the same thing. In Physics there is a very important difference and it is vital that you learn the above definitions.

It is also worth knowing that the gravitational field strength on the earth is 10N/kg and on the moon is 1.6N/kg.

The formula for weight is given below:

$$Weight = mass \; x \; gravitational \; field \; strength$$

Weight is measured in newtons, N.
Mass is measured in kilograms, kg.
Gravitational field strength is measured in newtons per kilogram, N/kg.

This formula can be written as:

$$W = m \; x \; g$$

Let's take a look at rearranging the formula. If we want to make **g** the subject of the formula, we need to divide both sides by **m**. This will give:

$$\frac{W}{m} = g$$

Therefore,

$$g = \frac{W}{m}$$

Going back to the original formula, $W = m \; x \; g$, if we want to make **m** the subject of the formula, we need to divide both sides by **g**. This will give:

$$\frac{W}{g} = m$$

Therefore,

$$m = \frac{W}{g}$$

You can now use the above formula to solve problems on weight. I will give 4 examples first before you have a go at the questions.

Example 1

Calculate the weight of a 20kg mass on earth. (The gravitational field strength on earth, g_earth =10N/kg)

m = 20kg
g = 10N/kg
W = ?

$$W = mg$$

$$W = 20 \times 10$$

$$W = 200N$$

Example 2

An object has a weight of 64N on the moon. Calculate it's weight on earth. (The gravitational field strength on the moon, g_moon = 1.6N/kg and g_earth = 10N/kg)

Firstly we need to calculate it's mass.

W = 64N
g = 1.6N/kg (remember it is on the moon!)
m = ?

$$W = mg$$

$$m = \frac{W}{g}$$

$$m = \frac{64}{1.6}$$

m = 40kg

Now we need to calculate the weight on earth.

m = 40kg
g = 10N/kg (remember it is on the earth!)
W = ?

$$W = mg$$

$$W = 40 \times 10$$

$$W = 400N$$

Example 3

An object on the surface of Jupiter weighs 300N and has a mass of 12kg. Calculate the gravitational field strength on Jupiter.
m = 12kg
W = 300N
g = ?

$$W = mg$$

$$g = \frac{W}{m}$$

$$g = \frac{300}{12}$$

$$g = 25 \text{N/kg}$$

Example 4

A ball has a mass of 500g. Calculate it's weight on earth. (g_{earth} = 10N/kg)

Firstly we must convert the 500g into kg!
To convert g into kg you divide by 1000.
Therefore 500g = 0.500kg.

m = 500g = 0.500kg
g = 10N/kg
W = ?

$$W = mg$$

$$W = 0.500 \times 10$$

$$W = 5.00\text{N}$$

Now it is time for you to attempt some questions.

Questions

1) Calculate the weight of the following masses on earth.
   a) 2kg
   b) 5kg
   c) 10kg
   d) 5.0×10^3kg
   e) 3.0×10^4kg
2) Calculate the mass of the following weights on the moon (g_{moon} = 1.6N/kg)
   a) 32N
   b) 80N
   c) 1.6×10^3N
   d) 6.4kN
   e) 3.2×10^4N
3) A jar of Jam has a mass of 400g. Calculate it's weight on:
   a) Earth
   b) Moon

4) Fully fueled at lift-off the Saturn V rocket had a weight of 2.8×10^7N. Calculate the mass of the rocket at lift-off (g_{earth}=10N/kg)
5) The Bugatti Veyron has a weight on earth of 18880N. The driver has a mass of 90kg. What is the combined mass of the car and driver on earth. (g_{earth}=10N/kg)

10. F = ma

The formula for the resultant force, F is given below:

$$Resultant\ force = mass\ x\ acceleration$$

Resultant force, F, is measured in newtons, N.
Mass is measured in kilograms, kg.
Acceleration is measured in metres per second squared, m/s².

If the mass remains constant, this formula is one way of expressing Newton's second law of motion.

Before we move on I need to explain what a resultant force is. The word 'resultant' means 'overall'. In this formula, it is extremely important to realise that the 'resultant' force is the 'overall' force. To further explain this see figures 1 and 2:

Figure 1

2N   10N

In figure 1, if you consider the toy tram and look at the forces acting on it. The 10N force to the right is being opposed by the 2N force to the left. Due to the fact that the 10N force is being opposed by the 2N force, to find the resultant force, we subtract the 2N force from the 10N force. Therefore:

$$resultant\ force = 10N - 2N$$

$$resultant\ force = 8N$$

(Strictly speaking, the size of the arrow should represent the size of the force. For example, a scale could be used so that every 2N in force is represented by an arrow of length 1cm. In figure 1, this would mean that the 10N force would be represented by an arrow 5cm long and the 2N force would be represented by an arrow 1cm long. However, when sketching diagrams and labelling the forces, it is okay just to draw the arrows approximately).

Let's consider another example.

Figure 2

In figure 2, the total force to the right = 8 + 4 =12N (the forces are added together because they are in the same direction). The total force to the left = 5N. The resultant force or overall force is calculated by:

$$resultant\ force = 12 - 5 = 7N$$

As you can see, with forces, it is very important to take the direction into account as well as the size or magnitude of the force (quantities where you have to take both the magnitude (size) and direction into account are known as vectors. Quantities where you only have to take the magnitude into account, for example in the case of speed, are known as scalars).

Let's go back and look at the formula again, which can be written as:

$$F = ma$$

Remember, F stands for the resultant force, not just force!
Let's take a look at rearranging the formula. If we want to make **a** the subject of the formula, we need to divide both sides by **m**. This will give:

$$\frac{F}{m} = a$$

Therefore,

$$a = \frac{F}{m}$$

Going back to the original formula, $F = ma$, if we want to make **m** the subject of the formula, we need to divide both sides by **a**. This will give:

$$\frac{F}{a} = m$$

Therefore,

$$m = \frac{F}{a}$$

You can now use the above formulae to solve problems using the formula F = ma. I will give 4 examples before you have a go at the questions.

Example 1

Using the toy tram in figure 1, calculate the acceleration of the tram if it has a mass of 4kg.

F = 10 – 2 = 8N (remember that **F** is the resultant force)
m = 4kg

a = ?

$$F = ma$$

$$a = \frac{F}{m}$$

$$a = \frac{8}{4}$$

a = 2 m/s²

If you go back and look at figure 1, there are 10N providing the driving force to move the tram forwards. In this example, the first 2N are used to overcome the 2N resistive force (the force to the left) and the remaining 8N are used to accelerate the tram!

Example 2

Calculate the resultant force acting on an object of mass 3kg which is accelerating at 2 m/s².

F = ?
a = 2 m/s²
m = 3kg

$$F = ma$$

$$F = 3 \times 2$$

F = 6N

Example 3

A model rocket of mass 10kg produces a thrust of 180N. Calculate the acceleration of the rocket. (g_{earth} = 10N/kg)

The first thing we need to do is to calculate the weight of the rocket.

W = ?
m = 10kg
g = 10N/kg

$$W = mg$$

$$W = 10 \times 10$$

W = 100N

It is often useful to draw a diagram, to show the forces acting on an object (in this case the rocket). See figure 3.

Figure 3

180N

100N

Note:
Weight is the force of gravity acting on an object and gravity acts towards the centre of the earth (or vertically down). Arrows showing the weight therefore need to be a straight line acting vertically down.

We can now use the diagram to help solve the problem.

F = 180 – 100 = 80N
m = 10kg
a = ?

$$F = ma$$

$$a = \frac{F}{m}$$

$$a = \frac{80}{10}$$

a = 8 m/s²

Example 4

A model car of mass 2.0kg accelerates at 1.0 m/s². Calculate the resultant force. If the drag force is 0.5N, calculate the force from the engine (Thrust).

(drag is the air resistive force acting to slow the car)

First we need to calculate the resultant force, F.

F = ?
m = 2.0kg
a = 1.0 m/s²

$$F = ma$$

$$F = 2.0 \times 1.0$$

$$F = 2.0N$$

For the next part it is worth drawing a diagram.

0.5N　　　　　　　　　　　　　　　　　　　Thrust

The resultant force will always be in the same direction as the acceleration. The car is accelerating forwards, so the 'Thrust' must be bigger than the drag force of 0.5N, for the resultant force to be in the forwards direction. The resultant force will therefore be equal to:

$$Resultant\ force = Thrust - 0.5$$

Firstly, we need to rearrange the formula for the **Thrust**.
To do this we need to add 0.5N to both sides of the formula. This gives:

$$Resultant\ force + 0.5 = Thrust$$

We have just calculated the resultant force to be 2.0N.
Therefore:

$$Thrust = resultant\ force + 0.5$$

$$Thrust = 2.0 + 0.5$$

$$Thrust = 2.5N$$

Questions

1) An object of mass 4kg has an acceleration of 3 m/s². Calculate the resultant force.
2) An object of mass 3kg has a resultant force acting on it of 18N. Calculate the acceleration.
3) A car accelerates at 2.500 m/s². If the mass of the car is 1500kg, calculate the resultant force acting on the car.
4) A Great Dane of mass 40kg accelerates at 2.0 m/s². Calculate the resultant force acting on the dog.
5) A car of mass 2.0×10^3kg accelerates at 3 m/s². If the drag force is 1000N, what is the thrust from the engine?
6) A model rocket has a mass of 8kg. If the thrust from the rocket is 120N, calculate the acceleration on the rocket.
7) The Saturn V rocket used by NASA for the Apollo missions had a mass of 2.80×10^6kg. If the thrust at launch was 34.5MN, calculate the acceleration.
8) An object has a resultant force acting on it of 3×10^4N. If it's

acceleration is 4.0 m/s², calculate the mass.
9) Santa has a sleigh which has a mass of 200kg when empty.
   a) Calculate the resultant force if it accelerates at 10 m/s².
   b) If the sleigh is then loaded with 300kg of presents, what resultant force would be required for the sleigh to have the same acceleration?
10) A 200kg motorbike has a resultant force acting on it of 1.0×10^3N.
    a) Calculate the acceleration.
    b) If the bike started from rest and the acceleration lasted for 5s, calculate the final velocity. (**Hint:** You will have to use another formula for the second part).

11. Momentum

Momentum is defined as the product of mass and velocity. The formula is given by:

$$momentum = mass \times velocity$$

momentum is measured in kilogram metres per second, kgm/s (this can be also written as kgms⁻¹). Another unit is the newton second, Ns. Both units are equivalent, so it doesn't matter which one you use.

Mass is measured in kilograms, kg.

Velocity is measured in metres per second, m/s.

The formula can be written as:
$$p = mv$$

The letter **p**, is used as the symbol for momentum. It is important to remember this.
Let's take a look at rearranging the formula.

$$p = mv$$

Taking a look at rearranging the formula, if we want to make **v** the subject of the formula, we need to divide both sides by **m**. This will give:

$$\frac{p}{m} = v$$

Therefore,
$$v = \frac{p}{m}$$

Going back to the original formula, $p = mv$, if we want to make **m** the subject of the formula, we need to divide both sides by **v**. This will give:

$$\frac{p}{v} = m$$

Therefore,
$$m = \frac{p}{v}$$

You can now use the above formulae to solve problems on momentum. I will give 4 examples before you have a go at the questions.

Example 1

A Steel ball of mass 2kg has a velocity of 10 m/s East. Calculate the momentum.

m = 2kg
v = 10 m/s
p = ?

$$p = mv$$

$$p = 2 \times 10$$

P = 20 kgm/s East

Note:
Momentum, like velocity is a vector and therefore has magnitude (size) and direction. This is why a direction has been given for the momentum.

Example 2

A car of mass 1.50×10^3kg has a velocity of 50.0km/h North. Calculate the momentum.

Firstly we need to convert 50.0km/h into m/s. We will do this in 2 stages. First will be to convert km/h into m/h (metres per hour). Then convert m/h into m/s. 'kilo' means 1000, so 50km/h = 50,000m/h. There are 3600s in 1 hour (60s = 1 minute and 60 minutes = 1 hour. 60 x 60 =3600s.) Therefore:

$$\frac{50,000}{3600} = 13.9$$

Therefore 50.0km/h = 13.9 m/s.
m = 1.50×10^3kg = 1500kg
v = 13.9 m/s
p = ?

$$p = mv$$

$$p = 1500 \times 13.9$$

P = 20,850 kgm/s

p = 20.9 x 10^3 kgm/s North

p = 20,900kgm/s North

(The answer must be to 3 significant figures because this was the minimum number given in the question. Therefore the answers in bold are your final correct answers. It is sometimes easier to leave the answer as a power of 10, as you can clearly see that **20.9 x 10^3 kgm/s** has 3 significant figures. With 20,900kgm/s it is unclear whether you have measured to the nearest 100kgm/s (therefore 3 significant figures) or to the nearest 1 kgm/s which would be 5 significant figures. Both answers in bold however, are correct and you can use either one.

Example 3

A car has a momentum of 6.0 x 10⁴ kgm/s South and a velocity of 30.0 m/s South. Calculate the mass of the car.

p = 6.0 x 10⁴ kgm/s = 60,000kg (see section 1 if you are unsure of this)
v = 30.0 m/s
m = ?

$$p = mv$$

$$m = \frac{p}{v}$$

$$m = \frac{60,000}{30.0}$$

m = 2000kg

Example 4

A ship has a momentum of 7.5 x 10⁵ kgm/s West and has a mass of 50,000kg. Calculate it's velocity.
p = 7.5 x 10⁵ kgm/s = 750,000 kgm/s
m = 50,000kg
v = ?

$$p = mv$$

$$v = \frac{p}{m}$$

$$v = \frac{750,000}{50,000}$$

v = 15 m/s West

Questions

1) A steel ball has a mass of 3.0kg. Calculate it's momentum if it has the following velocities:
   a) 2 m/s East   b) 4 m/s South   c) 10 m/s North   d) 4.0cm/s West
   e) 50mm/s North
2) An athlete has a mass of 100kg. Calculate his velocity if his momentum is:
   a) 500 kgm/s South   b) 1000 kgm/s East   c) 400 kgm/s West
3) A lorry of mass 10,000kg is travelling at 80.0km/h East. Calculate it's momentum.
4) A toy car of mass 100g is moving at 5cm/s North. Calculate it's momentum.
5) An athlete has a momentum of 600 Ns West. If she has a velocity of 8 m/s West, calculate her mass.

12. Force = rate of change of momentum

If we take a look at the formulae from section 8 & 10 we have:

$$a = \frac{v - u}{t}$$

And

$$F = ma$$

If we substitute the first formula into the second formula we get:

$$F = \frac{m(v - u)}{t} \quad (1)$$

This could also be written as:

$$F = m(v - u)/t$$

This formula can be expressed in a few different ways, but all are mathematically the same. Multiplying out the brackets we get:

$$F = \frac{(mv - mu)}{t} \quad (2)$$

Going back to formula (1) above, this can also be written as:

$$F = \frac{m}{t}(v - u)$$

We shall call this one formula (3).
Formulae (1), (2) and (3) are all mathematically the same, they are just written slightly differently.
Looking at formula (3) for example, this formula is useful in solving rocket problems, when the question gives you the mass of gas ejected from the rocket per second. See example 1 below:

Example 1

A rocket launched vertically ejects 40kg of gas each second. If the gas which is initially at rest reaches a velocity of 195 m/s, calculate the upward force on the rocket. If the mass of the rocket is 600kg, calculate the acceleration.

If we look at formula (3)

$$F = \frac{m}{t}(v - u)$$

The right hand side can be split into 2 sections, (m/t) and (v-u). (m/t) is mass divided by time which is like saying the mass of gas per second (in the question, this is 40kg/s). ($v - u$) is just the change in velocity (remember that **v** = final velocity and **u** = initial velocity). Substituting the numbers into the formula we get:

$$F = (40) \times (195 - 0)$$

$$F = 40 \times 195$$

$$F = 7800N$$

(The rocket exerts a force of 7800N on the exhaust gases. By Newton's third law of motion the exhaust gases exert an equal and opposite force on the rocket. **Therefore, the upwards force on the rocket is 7800N. Newton's third law states that if a body A exerts a force on body B, then body B exerts an equal and opposite force on A).**

The first thing that we need to do before we calculate the acceleration is to calculate the weight of the rocket.

m = 600kg
g = 10N/kg
W = ?

$$W = m \times g$$

$$W = 600 \times 10$$

$$W = 6000N$$

The rocket has an upwards force acting on it of 7800N and the weight of the rocket is 6000N (the weight acts vertically down). The resultant force acting on the rocket is therefore:

$$Resultant\ force, F = 7800 - 6000 = 1800N$$

Therefore,
m = 600kg
F = 1800N
a = ?

$$F = ma$$

$$a = \frac{F}{m}$$

$$a = \frac{1800}{600}$$

a = 3 m/s²

Example 2

A go-kart of mass 75kg changes its velocity from 2 m/s to 10 m/s in 4s. Calculate the force needed to cause this change in velocity.

For this example we will use formula (1):

$$F = \frac{m(v - u)}{t}$$

m = 75kg
v = 10 m/s
u = 2 m/s

t = 4s
F = ?

Substituting the numbers into the formula:

$$F = \frac{75(10-2)}{4}$$

$$F = \frac{(75 \times 8)}{4}$$

$$F = \frac{600}{4}$$

$$F = 150N$$

Let's now take a look at formula (2):

$$F = (mv - mu)/t$$

mv is the mass multiplied by the final velocity. Another name for this would be the final momentum. **mu** is the mass multiplied by the initial velocity. Another name for this would be the initial momentum. Therefore **mv – mu** is the final momentum – initial momentum. This is the change in momentum. Thus:

$$F = (mv - mu)/t$$

Can be written in words as:

Force = change in momentum / time.

(Force is equal to the change in momentum divided by the time)

This can also be stated as force equals the rate of change of momentum.

Also, if we look at the formula, F = (mv – mu)/t, multiplying both sides by **t**, we get:

$$Ft = \frac{(mv - mu)t}{t}$$

The **t**'s cancel on the right hand side to give:

$$Ft = mv - mu$$

Dividing both sides by **F** gives:

$$t = \frac{mv - mu}{F}$$

This then gives the time over which the momentum change has occurred.

Example 3

A ball changes its momentum from 6kgm/s to 20kgm/s. If a force of 7N caused this momentum change, for how long did this force act.

Initial momentum = 6kgm/s
Final momentum = 20kgm/s
Force = 7N
t = ?

$$F = \frac{mv - mu}{t}$$

$$t = \frac{mv - mu}{F}$$

$$t = \frac{20 - 6}{7}$$

$$t = \frac{14}{7}$$

t = 2s

Example 4

Calculate the force acting when an object has its momentum changed by 100kgm/s (or 100Ns) in 50ms.

Change in momentum = 100kgm/s
t = 50ms = 50 x 10^{-3}s = 0.050s
F = ?

$$Force = \frac{change\ in\ momentum}{time}$$

$$F = \frac{100}{0.050}$$

F = 2.0 x 10^3N

Or,

F = 2000N

Questions

1) Calculate the force acting when the momentum changes by:
   a) 100kgm/s in 5s.
   b) 20kgm/s in 0.4s
   c) 120kgm/s in 0.3s
   d) 800kgm/s in 0.2s
   e) 50kgm/s in 2s.
2) Calculate the change in momentum when:
   a) a 1000N force acts for 2s.
   b) a 3kN force acts for 0.4s.
   c) a 2 x 10^4N force acts for 5s.
   d) a 50N force acts for 100ms.

e) a 250N force acts for 0.5s.

Hint for Q2:

$$Ft = mv - mu$$

The change in momentum $(mv - mu)$ is equal to **Ft**. Therefore to calculate the change in momentum, all you need to do is multiply the force by the time.

3) A go-kart of mass 80kg changes its velocity from 1 m/s to 5 m/s in 4s. Calculate the force needed to cause this change in velocity.
4) A small rocket launched vertically ejects 50kg of mass of gas each second. The velocity of the gas changes from 0 m/s to 200 m/s.
   a) calculate the upward force acting on the rocket.
   b) If the mass of the rocket is 500kg, calculate the acceleration.
5) The momentum of an object changes from 100kgm/s to 400kgm/s. If the force acting is 50N, how long does the force act for in order to cause this change.

13. Conservation of momentum

The law of conservation of momentum states the following:
The total linear momentum of a system of interacting bodies (the bodies could be colliding or exploding, for example) on which no external forces (friction for example) are acting, remains constant.
This is a bit of a mouthful, and it is easier to write it as:
Momentum before = momentum after (unless external forces, for example friction, act).
This can be further reduced to:

$$p_{before} = p_{after} \text{ (unless external forces act)}$$

When solving problems on this topic it is important to know that momentum is a vector and as discussed in previous sections, direction is important. With momentum problems, it is usual to have objects moving to the right in diagrams to have a positive velocity and those moving to the left a negative velocity. If an object was moving upwards in a problem, it is usual to take the upwards direction to be positive and the downwards direction to be negative. Ultimately, you need to choose a direction to be positive.

Also, you need to try to solve the problems in 3 steps. These are as follows:
1) Calculate momentum before (p_{before})
2) Calculate momentum after (p_{after})
3) Apply the law of conservation of momentum ($p_{before} = p_{after}$) and solve.

Let's now take a look at a few examples in order to explain what I mean.

Example 1

Object A of mass 3kg travelling at 4 m/s to the right, collides with object B of mass 1kg which is stationary. On colliding they stick together. Calculate their velocity after collision.

It is useful when solving these problems to draw a diagram. Let's look at calculating the momentum before the collision (step 1). Firstly, remember that p = mv. To calculate the momentum before the collision, we need to add together the momentum of each object before the collision.

Figure 1 – Momentum before collision
(Right is positive!)

4 m/s →　　　　　　　　　0 m/s

A ◯ (3kg)　　　　　　◯ (1kg) B

Step 1 - p_before (remember, p = mv)

(3 x 4) + (1 x 0) = 12 + 0 = 12 kgm/s

Now let's look at the momentum after the collision.

Figure 2 – Momentum after collision

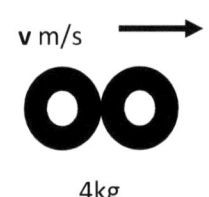

v m/s →

◯◯
4kg

Common sense will probably tell you that the objects will move to the right. Let's just assume that they do move to the right (if we are wrong, we will get a negative answer for velocity, which would show that they are actually moving to the left). We will call this velocity, **v**.

Step 2 - p_after (remember, p = mv)

p = (4 x **v**) = 4**v** kgm/s

Do not worry about the **v**. If **v** was 1 m/s, p = 4 kgm/s, if **v** was 2 m/s, p = 8 kgm/s. We will calculate the value of **v** in the next step!

Step 3 – By conservation of momentum, p_before = p_after

$$12 = 4v$$

To calculate **v**, divide both sides by 4.

$$\frac{12}{4} = \frac{4v}{4}$$

On the right hand side the 4's cancel, leaving:

$$3 = v$$

Therefore, **v** = 3 m/s right.

Example 2

A rifle has a mass of 5kg. If the rifle fires a bullet of mass 20g with a muzzle velocity of 500 m/s, calculate the velocity of recoil of the rifle.
(When a rifle is fired, it kicks back into the shoulder. This is known as the rifle recoiling)
Before looking at step 1, we need to ensure that everything is in the correct units, so we need to convert the mass into kilograms.

To convert g into kg you divide by 1000. Therefore, 20g = 0.020kg.

Step 1 – momentum before

Firstly, let's draw a diagram.
Figure 1 – momentum before

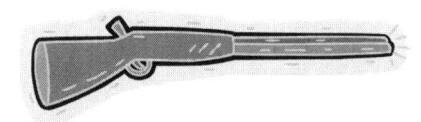

Initially, both the rifle and bullet are stationary (v = 0 m/s). Therefore, if we look at the momentum before, we have:

$$p = mv$$

$$(5 \times 0) + (0.020 \times 0) = 0 + 0 = 0 \text{ kgm/s}$$

Step 2 – momentum after

Again, let's draw a diagram.

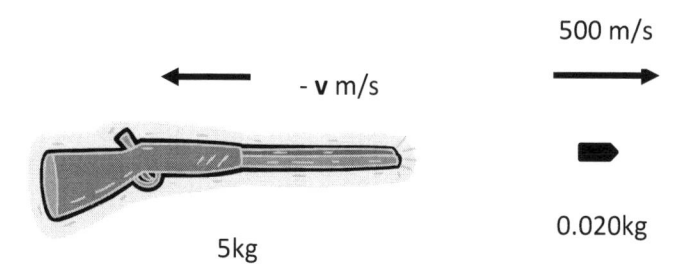

500 m/s

- v m/s

5kg

0.020kg

In this problem, we will again take the right to be positive. The bullet is moving to the right, so it's velocity will be + 500 m/s (there is no need to write the + sign though, as we know that 500 m/s is a positive number). Anything moving left will have a negative velocity. We do not know the velocity of the rifle, so I will call it **v**. Due to the rifle moving left, I will call it **– v** m/s (Do not be confused by this. If the rifle was moving left at 2 m/s we would be writing – 2 m/s (don't forget, the negative sign just indicates that it is moving left!) If the rifle was moving left at 5 m/s we would be writing – 5 m/s. We don't know what the velocity is, but we do know that it recoils in the opposite direction to the bullet, so we will call it **– v** m/s.

Note:
If you were not sure that the rifle was going to move to the left, this does not matter. Just write it down as a positive velocity (**v** m/s). If it turns out that it is moving to the left, you will just get a negative answer when you solve the problem. For example, - 2 m/s. The negative sign would just indicate that it is moving to the left.

Let's continue with the problem.

$$p = mv$$

$$(5 \times -v) + (0.020 \times 500)$$

(don't forget, we add the momentums together!)

This simplifies to:

$$-5v + 10$$

This is the momentum after. Now for step 3.

Step 3 - By conservation of momentum, $p_{before} = p_{after}$

$$0 = -5v + 10$$

If we add + 5v to both sides of the equation we get:

$$5v = -5v + 5v + 10$$

Therefore, $5v = 10$

(- 5v + 5v on the right hand side of the equation equals 0. This is no different to, for example, – 4 + 4 = 0 or – 7 + 7 = 0)

$$5v = 10$$

Dividing both sides by 5.

$$\frac{5v}{5} = \frac{10}{5}$$

The 5's cancel on the left hand side to give:

$$v = \frac{10}{5}$$

$$v = 2 \text{ m/s.}$$

Note:
We know that the rifle is moving left, because we have already taken this into account. If we hadn't known this and in the diagram we had the rifle moving to the right, our answer at the end of the calculation would have been – 2 m/s. The negative sign showing that the rifle moves to the left. It would be worth trying this for yourself, to aid your understanding of the topic.

Example 3

A rugby player of mass 70kg running north at 4.0 m/s collides with another player of mass 90kg running south at 6.0 m/s. If they cling together in the tackle, what will be their common velocity?

We will take the north to be positive and the south to be negative.

Step 1 – momentum before

$$p = mv$$

$$(70 \times 4.0) + (90 \times -6.0)$$

(don't forget, we add the momentums together)

$$\text{This equals, } 280 + -540$$

(don't forget also, that in Mathematics, $+ - = -$)

Therefore, we have:

$$280 - 540 = -260 \text{ kgm/s}$$

Step 2 – momentum after

Let's take both players afterwards, moving to the right, just to illustrate the final point of the last example, even though common sense should tell us that they will move left. We will have them moving with velocity **v**.

$$p = mv$$

Their combined mass will be $70 + 90 = 160$kg. Therefore:

$$p = 160 \times \mathbf{v}$$

$$p = 160v \text{ kgm/s}$$

This is the momentum after. Now for step 3.

Step 3 - By conservation of momentum, $p_{before} = p_{after}$

$$-260 = 160v$$

Dividing both sides by 160 we get:

$$-\frac{260}{160} = \frac{160v}{160}$$

$$-1.625 = v$$

Therefore, $v = -1.6$ m/s (or 1.6 m/s South!)

The answer needs to be given to 2 significant figures due to the smallest number of significant figures in the question being 2. Now it is your turn to have a go at some questions.

Questions

1) A rifle of mass 6.0kg fires a bullet of mass 15g. The rifle recoils at 2 m/s, calculate the velocity of the bullet.
2) An object of mass 2.0kg is travelling at 5.0 m/s and collides with a stationary object of mass 0.50kg. On colliding they stick together. With what velocity do they move?
3) A canoeist of mass 75kg is standing in his canoe of mass 25kg near to a river bank, both are stationary. The man steps out of the canoe onto the river bank and the canoe goes in the opposite direction at 2.0 m/s. Calculate the velocity of the canoeist.
4) A car of mass 2000kg is travelling East at 20.0 m/s. It collides with a lorry of mass 10,000kg travelling West at 30.0 m/s. On colliding, they stick together. What is their velocity after impact?
5) A wooden block of mass 1.98kg is hanging from a string. The wooden block is stationary. A bullet of mass 0.020kg travelling at 600 m/s hits the block and embeds into it. With what velocity do the 'block and bullet' move?

14. Work done

The formula for Work done is given below:

Work done = Force x distance moved in the direction of the force

Work done is measured in joules, J.
Force is measured in newtons, N.
Distance moved in the direction of the force is measured in metres, m.

This formula can be written as:

$$W = F \times d$$

Note:
Be careful that you don't get mixed up with the formula that we looked at in section 9, $W = mg$, where the **W** stands for weight!

Let's take a look at rearranging the formula. If we want to make **F** the subject of the formula, we need to divide both sides by **d**.

$$\frac{W}{d} = F$$

Therefore,
$$F = \frac{W}{d}$$

If we now look at making **d** the subject of the formula, going back to the original formula, $W = F \times d$, we need to divide both sides by **F**. We will then get:

$$\frac{W}{F} = d$$

Therefore,

$$d = \frac{W}{F}$$

Before I give some examples on this formula, there is a very important statement which you need to know about. It is stated as:

Work done = energy transferred

Both 'Work done' and 'energy transferred, are measured in joules, J. We will come back to this statement as and when we need it. Now back to the examples.

Example 1

I am a former British Bench Press Champion with a personal best in competition of 152.5kg. Let's do an example of me bench pressing.

How much work is done in bench pressing 152.5kg from chest to arms length (50.0cm).

If we look at the formula:

$$W = F \times d$$

We have a distance, but we have been given the mass. We need to convert the mass to a weight (remember that weight is a force!)

g_{earth} = 10N/kg
m = 152.5kg
W = ?

$$W = mg$$

$$W = 152.5 \times 10$$

$$W = 1525N$$

Therefore,

F = 1525N (again, weight is a force!)
d = 50cm = 50 x 10^{-2}m = 0.50m (To convert cm to m you could also divide by 100, but remember that distance must be measured in metres!)
W = ?

$$W = 1525 \times 0.50$$

$$W = 762.5N$$
$$W = 763N$$

(remember what we have mentioned about significant figures and rounding)

Example 2

Figure 1

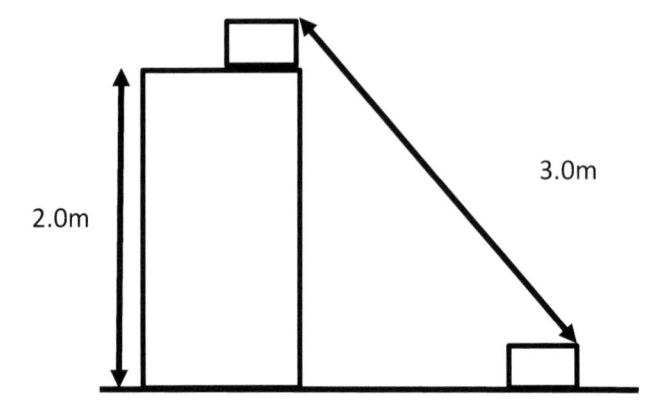

A box of weight 20N is lifted from the floor onto the top of a cupboard as shown in figure 1. How much work is done?

Looking at this problem, we are given 2 distances, 2.0m & 3.0m. Which distance do we use? We need to go back to the formula to explain which distance to choose:

Work done = Force x distance moved in the direction of the force

It is the 'distance moved in the direction of the force' which is important here. In lifting the box, you are working against gravity. Gravity acts towards the centre of the earth or vertically down. We must therefore, in this example, use the vertical distance!

F = 20N
d = 2.0m
W = ?

$$W = F \times d$$

$$W = 20 \times 2.0$$

$$W = 40J$$

Example 3

An object is dragged up a slope, as shown in figure 2, by a pull force of 50N. How much work is done by this force?

Looking at this problem, we are given 2 distances again, 2.0m & 3.0m. Which distance do we use? Again, we need to go back to the formula to explain which distance to choose:

Figure 2

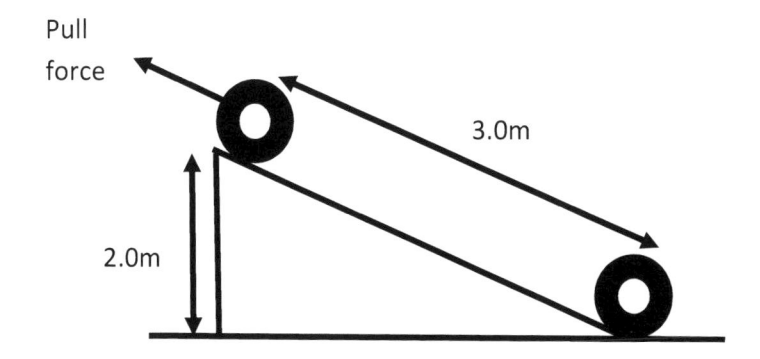

Pull force

2.0m

3.0m

Work done = Force x distance moved in the direction of the force

It is the 'distance moved in the direction of the force' which is the important point again. The box is being pulled up the slope and we must use the distance in the direction of the force. In this case, we must use 3.0m as the distance!

F = 50N
d = 3.0m
W = ?

$$W = F \times d$$

W = 50 x 3.0

W = 150J

Example 4

A crane does 45,000J of work when lifting a 3.0kN weight. Through what vertical height is it lifted?

W = 45,000J
F = 3kN = 3 x 10^3N = 3000N (remember force must be in newtons and weight is a force)
d = ?

$$W = F \times d$$

$$d = \frac{W}{F}$$

$$d = \frac{45,000}{3000}$$

d = 15m

Questions

1) A powerlifter bench presses 140kg from chest to arms length

(60.0cm). How much work is done?
2) A 5.0kg box is lifted from the floor onto a shelf 2.5m high. How much work is done?
3) A box is dragged along the floor by a rope with a pull force of 70N. If the box is dragged 10m, how much work is done?
4) A 3.0kg box is lifted from the floor onto a shelf 2000mm high. How much work is done?
5) A crane does 20,000J of work when lifting a 200kg mass. Through what vertical height is it lifted?
6) A book is lifted from the floor onto a book shelf through a distance of 150cm. If 7.5J of work are done, what is the **mass** of the book?
7) The world's strongest man deadlifts a mass of 425kg. He does 3825J of work. Through what vertical distance does it move?
8) The world's strongest man in the previous question did 3825J of work while deadlifting. How much energy did he transfer while doing this?

15. Kinetic energy

Kinetic energy is the energy an object has due to its movement.
The formula for kinetic energy is given below:

$$\text{Kinetic energy} = \tfrac{1}{2} \times \text{mass} \times \text{speed}^2$$

kinetic energy is measured in joules, J.
Mass is measured in kilograms, kg.
Speed is measured in metres per second, m/s.

This formula can be written as:

$$E_k = \tfrac{1}{2} \times m \times v^2$$

Where E_k is the kinetic energy, **m** is the mass and **v** is the speed.
Which is the same as:

$$E_k = \tfrac{1}{2} m v^2$$

Let's take a look first at rearranging the formula.

$$E_k = \tfrac{1}{2} m v^2$$

If we want to make **m** the subject of the formula (or rearranging for **m**), firstly we need to **multiply both sides by 2**, in order to remove the 2 from the denominator position on the right hand side of the formula. This gives:

$$2E_k = mv^2$$

Now dividing both sides by v^2, we get:

$$\frac{2E_k}{v^2} = m$$

Therefore,

$$m = \frac{2E_k}{v^2}$$

(This could also be written as **m = 2E$_k$ /v²**)

In order to make **v** the subject of the formula, we need to make **v²** the subject of the formula first and then square root this to find **v**.

$$E_k = \tfrac{1}{2} mv^2$$

Multiplying both sides by 2 again, we get:

$$2E_k = mv^2$$

Now to get **v²** by itself, we divide both sides by **m**, giving:

$$\frac{2E_k}{m} = v^2$$

Therefore, square rooting both sides, we get:

$$v = \sqrt{\frac{2E_k}{m}}$$

Note:
If you prefer, there is another way to rearrange the kinetic energy formula which I will show below. You can then choose the method that you prefer.

Method 2

$$E_k = \tfrac{1}{2} mv^2$$

If we look at the right hand side of the formula, the **½**, **m** and **v²** are all multiplied together. Therefore, if we want to make **m** subject of the formula, we need to divide both sides by **(½ v²)**. This will give:

$$\frac{E_k}{(\tfrac{1}{2} \times v^2)} = m$$

Therefore,

$$m = \frac{E_k}{(\tfrac{1}{2} \times v^2)}$$

(remember to work out the brackets first! BIDMAS!)

Again, if we look at the kinetic energy formula, **E$_k$ = ½ mv²**, the right hand side of the formula shows that the **½**, **m** and **v²** are all multiplied together. Therefore, if we want to make **v²** the subject of the formula, we need to divide both sides by **(½ m)**. This will give:

$$\frac{E_k}{(\tfrac{1}{2} \times m)} = v^2$$

Therefore,

$$v^2 = \frac{E_k}{(\tfrac{1}{2} \times m)}$$

(remember that this is the same as $v^2 = E_k / (½ \times m)$)

If we then square root both sides, we get:

$$v = \sqrt{E_k / (½ \times m)}$$

My advice on this one is to work the calculation out in stages to avoid any calculator mistakes.

1) Calculate what is inside the brackets (½ x m).
2) Divide **E_k** by the answer inside the brackets.
3) Square root your final answer from stage 2.

You now have 2 methods for rearranging this formula and you can choose whichever you prefer, although ideally, you will be happy with both methods.

Example 1

An object of mass 4kg has a speed of 10 m/s. Calculate it's kinetic energy.

m = 4kg
v = 10 m/s
E_k = ?

$$E_k = ½ mv^2$$

$$E_k = ½ \times 4 \times 10^2$$

(remember that $10^2 = 10 \times 10 = 100$ and also remember that you must square the 10!)

$$E_k = ½ \times 4 \times 100$$

$$E_k = 200J$$

Example 2

A motorbike has a kinetic energy of 202,500J. If it has a mass of 200kg, calculate it's speed.

E_k = 202,500J
m = 200kg
v = ?

$$E_k = ½ mv^2$$

$$v^2 = \frac{2E_k}{m}$$

$$v^2 = \frac{2 \times 202,500}{200}$$

$$v^2 = \frac{405,000}{200}$$

$v^2 = 2025$ (square root both sides)

(If you square root v^2, you just get v).

$$v = \sqrt{2025}$$

$$v = 45 \text{ m/s.}$$

Note:
You can check that you have got this correct by putting the answer back into the kinetic energy formula and see if you get a kinetic energy of 202,500J.

$$E_k = \tfrac{1}{2} mv^2$$

$$E_k = \tfrac{1}{2} \times 200 \times 45^2$$

$$E_k = \tfrac{1}{2} \times 200 \times 2025$$

$$E_k = 202,500 \text{J}$$

This is what we were given in the question, so we know that we are correct.

Example 3

An object has a kinetic energy of 1.60×10^4J and a speed of 40 m/s. Calculate it's mass.

$E_k = 1.60 \times 10^4 \text{J} = 16,000 \text{J}$
$v = 40$ m/s
$m = ?$

$$E_k = \tfrac{1}{2} mv^2$$

$$m = \frac{2E_k}{v^2}$$

$$m = \frac{2 \times 16,000}{40^2}$$

$$m = \frac{32,000}{1600}$$

$$m = 20 \text{kg}$$

Questions

1) An steel block of mass 8.0kg has a speed of 5.0 m/s. Calculate it's kinetic energy.
2) A golf ball of mass 45.0g has a speed of 30.0 m/s. Calculate it's kinetic energy.
3) A Javelin of mass 800g has a speed of 35.0 m/s. Calculate it's kinetic energy.
4) A car has a kinetic energy of 2.5MJ and a speed of 50 m/s. Calculate it's mass
5) A fish has a kinetic energy of 9.0J. If it has a mass of 2.0kg, calculate it's speed.
6) An object of mass 15.0kg has a kinetic energy of 3.0kJ. Calculate it's

speed.
7) The Saturn V rocket, 20s after launch had an altitude of 510m and a speed of 56 m/s. If it's mass is 2.8 x 10^6kg, calculate it's kinetic energy.
8) Thrust SSC (supersonic car) has a mass of 10.5 x 10^3kg reached a top speed of 763mph (miles per hour). Calculate it's kinetic energy. (1 mile = 1609m)
9) A car has a mass of 1500kg and a kinetic energy of 216,750J. Calculate its speed.
10) A lorry travelling at 26m/s has a kinetic energy of 3.211MJ. Calculate its mass.

16. Gravitational potential energy

Gravitational potential energy is the energy that a body has due to its position.

The formula for gravitational potential energy is given by:

Change in gravitational potential energy = mass x gravitational field strength x change in height

Change in gravitational potential energy is measured in joules, J.
Mass is measured in kilograms, kg.
Gravitational field strength is measured in newtons per kilogram, N/kg.
Change in height is measured in metres, m.

This formula can be written as:
$$E_p = m \times g \times h$$
Or,
$$E_p = mgh$$

Let's take a look at rearranging the formula. If we look at the right hand side of the formula the **m, g** and **h** are all multiplied together. If we want to make **m** the subject of the formula, we need to divide both sides by **gh**. This gives:

$$\frac{E_p}{(gh)} = m$$

Therefore,
$$m = \frac{E_p}{(gh)}$$

Note:
It is worth putting the **gh** in brackets because this will avoid any possible calculator mistakes. Using **BIDMAS** you will then calculate what is inside the brackets first and then divide E_p by the answer calculated from inside the brackets.

Again, if we look at the gravitational potential energy formula, **E_p = mgh**, the right hand side of the formula shows that the **m, g** and **h** are all multiplied together. If we want to make **g** the subject of the formula, we need to divide both sides by **mh**. This gives:

$$\frac{E_p}{(mh)} = g$$

Therefore,

Physics calculations for GCSE & IGCSE

$$g = \frac{E_p}{(mh)}$$

I have put the **mh** in brackets for the reason mentioned above.

Again, if we look at the gravitational potential energy formula, E_p = **mgh**, the right hand side of the formula shows that the **m, g** and **h** are all multiplied together. If we want to make **h** the subject of the formula, we need to divide both sides by **mg**. This gives:

$$\frac{E_p}{(mg)} = h$$

Therefore,

$$h = \frac{E_p}{(mg)}$$

Please note again the use of brackets.

If we look again at the formula for E_p, E_p = **mgh**. Notice that on the right hand side of the formula we have **mg**. Going back to section 9, Weight, W = mg. We can therefore replace in the E_p formula, **mg** for **W**. This will give another useful formula for E_p.

$$E_p = W \times h$$

Or,

$$E_p = Wh$$

Therefore E_p = weight x change in height, where weight is measured in newtons, N.

Note:
If we look at the right hand side of this formula, we have 'weight x change in height'. Weight is a force and height is a distance. This is the same as what 'work done' is equal to!

$$W = F \times d$$

If you remember though from section 14, we also have:

Work done = energy transferred

Therefore, the work done in lifting an object, is equal to its gain in gravitational potential energy.

Let's take a look at rearranging this formula:

$$E_p = Wh$$

To make **W** the subject of the formula, we need to divide both sides by **h**. This gives:

$$\frac{E_p}{h} = W$$

Therefore,

$$W = \frac{E_p}{h}$$

Looking back at the gravitational potential energy formula, $E_p = Wh$, to make **h** the subject of the formula, we need to divide both sides by **W**. This gives:

$$\frac{E_p}{W} = h$$

Therefore,

$$h = \frac{E_p}{W}$$

My advice when solving problems on gravitational potential energy is to use $E_p = mgh$ if you are given the mass in the question and use $E_p = Wh$ if you are given the weight in the question. Can you see from this the importance of knowing the difference between mass and weight!

Another important point, is that when objects are being raised in height, you are working against gravity and so **vertical distances must be used**. You may need to read again the 'work done' section.

Example 1

A climber of mass 75kg climbs to the top of a mountain, 4000m above where he started from. Calculate the gain in gravitational potential energy.
($g_{earth} = 10N/kg$)

m = 75kg
g = 10N/kg
h = 4000m

$$E_p = mgh$$

$$E_p = 75 \times 10 \times 4000$$

$$E_p = 3,000,000J$$

Or,

$$E_p = 3.0 \times 10^6 J$$

Example 2

A Boeing 737 has increased its gravitational potential energy since take-off by $6.3 \times 10^9 J$. If it has changed its height by 9km, what is the mass of the aircraft?
($g_{earth} = 10N/kg$)

$E_p = 6.3 \times 10^9 J = 6,300,000,000J$
$h = 9km = 9 \times 10^3 m = 9000m$
$g = 10N/kg$
$m = ?$

$$E_p = mgh$$

$$m = \frac{E_p}{(gh)}$$

Physics calculations for GCSE & IGCSE

$$m = \frac{6{,}300{,}000{,}000}{(10 \times 9000)}$$

working out the brackets out first we get:

$$m = \frac{6{,}300{,}000{,}000}{90{,}000}$$

$$m = 70{,}000 \text{kg}$$

Example 3

An object gains 600J of gravitational potential energy when lifted through a height of 30m. Calculate a) Weight b) mass of the object.

$E_p = 600J$
$h = 30m$
$W = ?$

$$E_p = Wh$$

$$W = \frac{E_p}{h}$$

$$W = \frac{600}{30}$$

$$W = 20N$$

To calculate the mass, $W = mg$

$$m = \frac{W}{g}$$

$$m = \frac{20}{10}$$

$$m = 2 \text{kg}$$

Questions (g_{earth} = 10N/kg)

1) Calculate the change in gravitational potential energy for a 5.0kg mass if the height changes by:
   a) 10m b) 25m c) 1000m d) 2km e) 3×10^3m
2) Calculate the change in height if a 20kg mass has its gravitational potential energy changed by:
   a) 1000J b) 3000J c) 7000J d) 8kJ e) 5.0×10^5J
3) When a crane lifts a 20,000N weight its gravitational potential energy changes by 0.60MJ. Calculate the change in height.
4) Spider-man is climbing up a skyscraper. Initially he is 20m above the ground. In climbing to 50m above the ground his gravitational potential energy changes by 24,000J. Calculate a) His weight. b) His mass
5) The Saturn V rocket, 20s after launch had an altitude of 510m and a speed of 56 m/s. If the mass of the rocket is 2.8×10^6kg, calculate the

change in gravitational potential energy.
6) A golf ball of mass 45g has its gravitational potential energy changed by 13.5J. Calculate the change in height.
7) A 6N weight is lifted through a vertical distance of 2500mm. Calculate its change in gravitational potential energy.
8) A group of Physics students have made a rocket that they want to test. The rocket has a mass of 3.0kg. If the change in gravitational potential energy is 1.2×10^3J, calculate the change in height.

17. Conservation of energy

Conservation of energy states that energy may be transformed from one form to another, but can't be created or destroyed, i.e. the total energy of a system remains constant. We will look at a particular case of gravitational potential energy transforming into kinetic energy and vice versa.
Let's initially take a look at a ball on top of a shelf.
If the ball has a mass of 1kg and is placed on top of a shelf 1m high, then if g = 10N/kg, the gravitational potential energy, E_P, will equal 10J (remember from the last section that $E_P = m \times g \times h$). If the ball is stationary, then its initial kinetic energy, E_k, will be 0J ($E_k = \frac{1}{2}mv^2$ and **v** = 0m/s). The ball is then 'just' knocked off the shelf. See figure 1 below.
(In all of these problems, we will ignore air resistance).

Figure 1

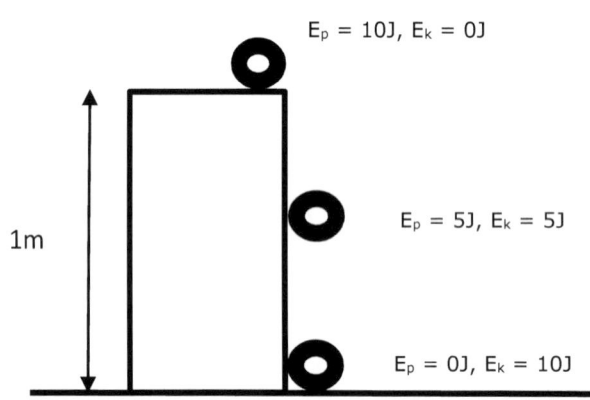

$E_p = 10J, E_k = 0J$

$E_p = 5J, E_k = 5J$

1m

$E_p = 0J, E_k = 10J$

As we have just shown, the ball initially has 10J of E_P and 0J of E_k (importantly, the total energy is 10J).

Half way down the ball will have an E_P of 5J (this can be calculated from $E_P = mgh$, with m = 1kg, g = 10N/kg and h = 0.5m (thinking at a slightly higher level, from the formula you could say that E_P was directly proportional to the height ($E_P \propto h$). Therefore, if the height has halved, the E_P would halve). Also, at the half way point, the ball would have gained 5J of E_k. Just before the instant of impact the E_P is now 0J (this is because h = 0m and $E_P = mgh$) and the E_k is 10J. Can you see that at any point, the total energy is the same, 10J. Energy has been conserved! It has just been transformed into another form. The E_p lost is equal to the E_k gained!

Example 1

A ball of mass 2.0kg is lifted a height of 3.0m above the ground. It is then released. Calculate: a) its gravitational potential energy b) its kinetic energy when it hits the ground c) its speed on impact with the ground d) its gravitational potential energy if it rebounds to a height of 2.0m e) its kinetic energy when it left the ground. (g_{earth} = 10N/kg)

a) m = 2.0kg
    g = 10N/kg
    h = 3.0m

$$E_p = mgh$$

$$E_p = 2.0 \times 10 \times 3.0$$

$$E_p = 60J$$

b) By conservation of energy, E_p lost = E_k gained, therefore, the kinetic energy when it hits the ground = 60J (You must use conservation of energy in this situation. You can't use the kinetic energy formula, as you do not know the speed or kinetic energy on impact yet).

c) E_k = 60J
    m = 2.0kg
    v = ?

$$E_k = \tfrac{1}{2} mv^2$$

from section 15,

$$v^2 = \frac{E_k}{(\tfrac{1}{2} \times m)}$$

$$v^2 = \frac{60}{(\tfrac{1}{2} \times 2.0)}$$

$$v^2 = \frac{60}{1.0}$$

$$v^2 = 60$$

(now if we square root both sides)

$$v = 7.746 \text{ m/s}$$

$$v = 7.7 \text{ m/s}$$

d) m = 2.0kg
    g = 10N/kg
    h = 2.0m

$$E_p = mgh$$

$$E_p = 2.0 \times 10 \times 2.0$$

$$E_p = 40J$$

Note:
This is less than the original 60J, but don't be fooled, energy has still been conserved. 20J have been transformed to heat and sound energy on impact with the ground. 40 + 20 = 60J. Energy has been conserved!

e) By conservation of energy, E_k lost = E_p gained. Therefore, if the ball has an E_p of 40J, it must have had 40J of kinetic energy initially on leaving the ground. The 40J of kinetic energy, was transformed to 40J of gravitational potential energy. The answer is therefore 40J.

Questions (ignore air resistance for all questions and g_{earth} = 10N/kg)

1) A ball of mass 3.0kg is lifted a height of 4.0m above the ground. It is then released. Calculate: a) its gravitational potential energy b) its kinetic energy when it hits the ground c) its speed on impact with the ground d) its gravitational potential energy if it rebounds to a height of 3.0m e) its kinetic energy when it left the ground.
2) An object has a gravitational potential energy of 50J, which has been calculated from ground level. What is the kinetic energy as it hits the ground.
3) A object of mass 5.00kg is lifted a height of 10.0m above the ground. It is then released. Calculate: a) its gravitational potential energy b) its kinetic energy when it hits the ground c) its speed on impact with the ground d) its gravitational potential energy if it rebounds to a height of 8.0m e) its speed when it left the ground.
4) A ball of mass 2.0kg is thrown vertically upwards with a speed of 5.0 m/s. Calculate a) kinetic energy b) the height reached
5) A parachutist of mass 100kg jumps from an aeroplane at a height of 3000m. In falling to a height of 2600m, calculate a) the loss in gravitational potential energy b) the gain in kinetic energy c) the speed reached at 2600m.

18. Moments (or Torques)

A 'moment' is 'the turning effect of a force'. The formula for a moment is given below:

Moment = Force x perpendicular distance from the pivot

A 'moment' is measured in newton-metres, Nm.
Force is measured in newtons, N.
Perpendicular distance from the pivot is measured in metres, m.

Note:
'Perpendicular' means at a right angle (90°).

This formula can be written as:

$$M = F \times d$$

Let's take a look at rearranging the formula. To make **F** the subject of the formula, divide both sides by **d**. This will give:

$$\frac{M}{d} = F$$

Therefore,
$$F = \frac{M}{d}$$

If we now look at the original formula again (M = F x d), to make **d** the subject of the formula, we need to divide both sides by F. This will give:

$$\frac{M}{F} = d$$

Therefore,
$$d = \frac{M}{F}$$

Example 1

A 20N force is applied to a spanner. The perpendicular distance of the applied force from the pivot is 20cm. Calculate the moment.

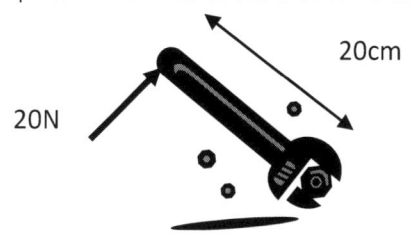

F = 20N
d = 20cm = 20 x 10^{-2}m = 0.20m (To convert cm to m you could also divide by 100, but remember that **d** must be measured in metres!)
M = ?

$$M = F \times d$$

$$M = 20 \times 0.20$$

$$M = 4Nm$$

A moment (or torque) is a vector. This means that it has magnitude (size) and direction. Strictly speaking therefore, a direction is needed. For moments, the directions used are **clockwise** or **anticlockwise**. If you look at the force acting on the spanner, it causes the spanner to turn clockwise. Therefore the answer is:

$$M = 4Nm \ \ clockwise$$

Example 2

A force of 100N is applied to a crow bar. If the moment is 50Nm, calculate the perpendicular distance of the force from the pivot.

F = 100N
M = 50Nm
d = ?

$$M = F \times d$$

$$d = \frac{M}{F}$$

$$d = \frac{50}{100}$$

$$d = 0.50m$$

Example 3

A door has a force applied to it. If the moment caused by the force is 20.1Nm and the perpendicular distance of the force from the pivot is 0.67m, calculate the force.

M = 20.1Nm
d = 0.67m
F = ?

$$M = F \times d$$

$$F = \frac{M}{d}$$

$$F = \frac{20.1}{0.67}$$

$$F = 30N$$

Questions

1) The perpendicular distance of the applied force to a spanner is 15cm from the pivot. Calculate the moment if the force applied is:
   a) 5.0N b) 10N c) 12N d) 20N e) 30N
2) A force of 200N is applied to a crow bar. Calculate the perpendicular distance of the force from the pivot if the moment is:
   a) 400Nm b) 300Nm c) 200Nm d) 150Nm e) 100Nm
3) A door has a force applied to it. If the moment caused by the force is 16Nm and the perpendicular distance of the force from the pivot is 0.80m, calculate the force.
4) Figure 1

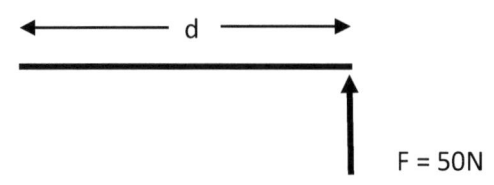

In figure 1 above diagram calculate **d** if the moment is:
   a) 500Nm anticlockwise b) 250Nm anticlockwise c) 25Nm anticlockwise d) 75Nm anticlockwise e) 150Nm anticlockwise
5) A 16N force is applied to a spanner. The perpendicular distance of the applied force from the pivot is 150mm. Calculate the moment.

Physics calculations for GCSE & IGCSE

19. Law of moments

The law of moments states the following:

For equilibrium (or for balance)

    Sum of the clockwise moments = Sum of the anticlockwise moments

I will give some examples to explain this law.

Example 1

Figure 1

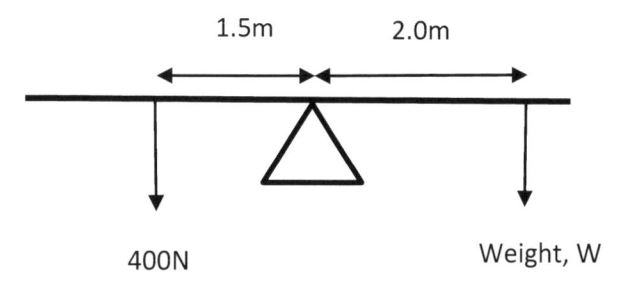

1.5m   2.0m

400N   Weight, W

Figure 1 above shows a see-saw. Connor of weight 400N sits 1.5m away from the pivot, while Libby sits the other side of the pivot 2.0m away. Calculate the weight of Libby if the see-saw is balanced (in equilibrium).
Note:
It is important to remember that weight is a force!

Firstly, if we look at the 400N force, this causes the see-saw to move down to the left of the pivot. As discussed in the 'moments' section though, we need to use 'clockwise' or 'anticlockwise', **not** 'up', 'down', 'left' or 'right'. Therefore, just considering the 400N force, this causes the see-saw to move 'anticlockwise'

The first thing we need to do then is calculate the 'anticlockwise' moment.

Anticlockwise moment

$$M = F \times d$$

$$M = 400 \times 1.5$$

$$M = 600 \text{ Nm anticlockwise}$$

Next, we need to calculate the 'clockwise' moment. Libby is on the right hand side of the pivot and just looking at the weight of Libby, this will cause the see-saw to move down to the right. Sticking to our convention, this causes the see-saw to move 'clockwise'.

Clockwise moment

$$M = F \times d$$

$$M = W \times 2.0 \quad \text{(remembering that weight is a force!)}$$

$$M = 2.0W \text{ Nm}$$

Do not be put off here by the fact that we do not have a single number. We will calculate W in the next step.
The final step is to apply the 'law of moments' and put the 'clockwise' moment equal to the 'anticlockwise' moment.

Applying the law of moments

Sum of the clockwise moments = Sum of the anticlockwise moments

Therefore,
$$2.0W = 600$$

If we want to make **W** the subject of the formula, we need to divide both sides by 2. This gives:

$$W = \frac{600}{2.0}$$

$$W = 300N$$

When solving problems of this kind there are 3 steps to the problem.

Step 1 – calculate the 'anticlockwise' moment
Step 2 – calculate the 'clockwise' moment
Step 3 – Apply the 'law of moments' and solve

(steps 1 & 2 can be done in any order)

Example 2

Figure 2

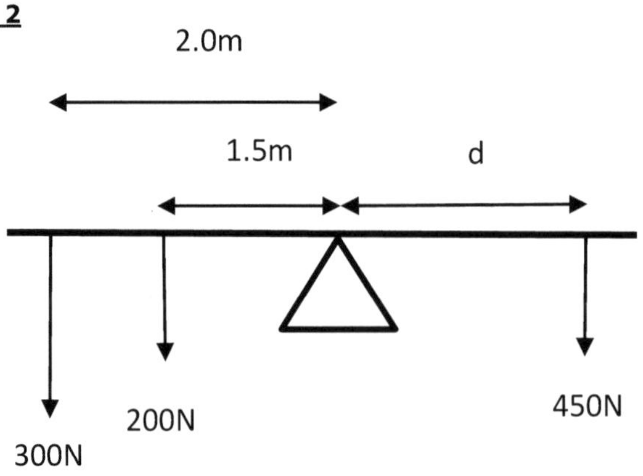

Using figure 2, calculate distance **d** if the see-saw is in equilibrium (balanced).

If we look at the 3 forces individually, the 200N and 300N forces cause the see-saw to move 'anticlockwise' and the 450N force causes it to move 'clockwise'. The first thing to do is calculate the **total** 'anticlockwise' moment.

This is done by calculating the 'moments' due to the 300N and 200N forces and adding them together.

Step 1 - Anticlockwise moment

$$M = F \times d$$

$$(300 \times 2.0) + (200 \times 1.5)$$

$$600 + 300$$

$$900 Nm$$

Step 2 - Clockwise moment

$$M = F \times d$$

$$M = 450 \times d$$

$$M = 450d$$

Step 3 - Applying the law of moments

Sum of the clockwise moments = Sum of the anticlockwise moments

Therefore,
$$450d = 900$$

To make **d** the subject of the formula, we need to divide both sides by 450. This gives:

$$d = \frac{900}{450}$$

$$d = 2.0m$$

Example 3

Figure 3

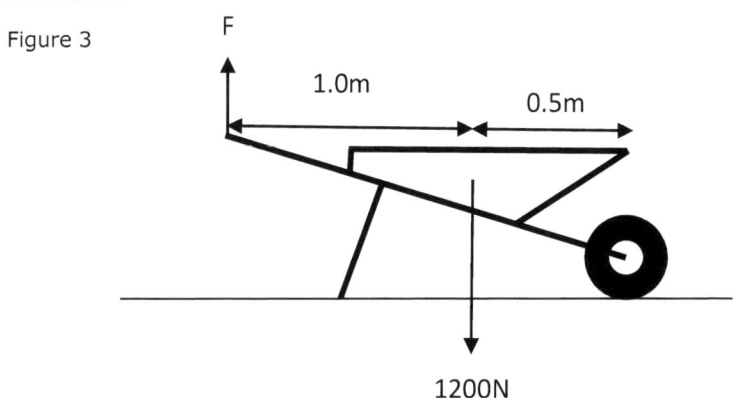

1200N

Figure 3 shows a diagram of a wheelbarrow. The pivot is at the centre of the wheel and the weight of the wheelbarrow and contents is 1200N. Calculate the force **F**, which is needed to just lift the wheelbarrow off the ground.

Looking at the 1200N force only, this will cause the wheelbarrow to rotate (turn) anticlockwise. The force **F**, needed to just lift the wheelbarrow off the ground will cause the wheelbarrow to rotate clockwise. Let's now follow the 3 steps.

Step 1 - Anticlockwise moment

$$M = F \times d$$
$$M = 1200 \times 0.5$$
$$M = 600 Nm$$

Step 2 - Clockwise moment

$$M = F \times d$$
$$M = F \times (1.0 + 0.5)$$

It is important to remember that 'Moment = Force x perpendicular distance from the pivot'. We therefore need to add 1.0m and 0.5m together to find the perpendicular distance from the pivot of force **F**.

$$M = F \times 1.5$$
$$M = 1.5F$$

Step 3 - Applying the law of moments

Sum of the clockwise moments = Sum of the anticlockwise moments

$$1.5F = 600$$

To make **F** the subject of the formula, we need to divide both sides by 1.5. This gives:

$$F = \frac{600}{1.5}$$

$$F = 400N$$

Questions

1) Figure 4 shows a see-saw in equilibrium. Calculate the weight, **W**.

Figure 4

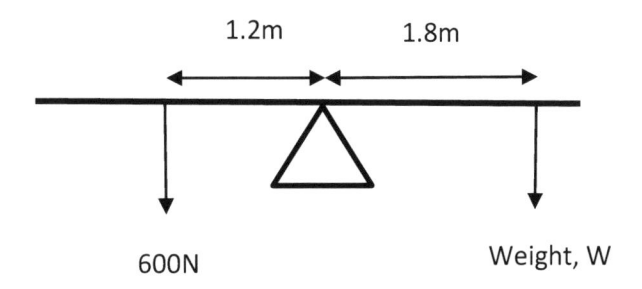

```
        1.2m      1.8m
   ←——————→←——————→
  ━━━━━━━━━━━━━━━━━━━━━━
            △
   ↓                  ↓
  600N           Weight, W
```

2)

Figure 5

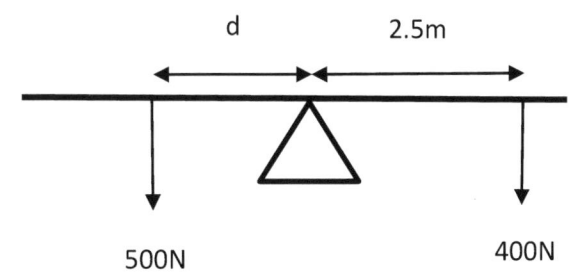

```
        d         2.5m
   ←————→←——————————→
  ━━━━━━━━━━━━━━━━━━━━━━
            △
   ↓                  ↓
  500N              400N
```

Figure 5 shows a see-saw in equilibrium. Calculate the distance, **d**.

3) Figure 6

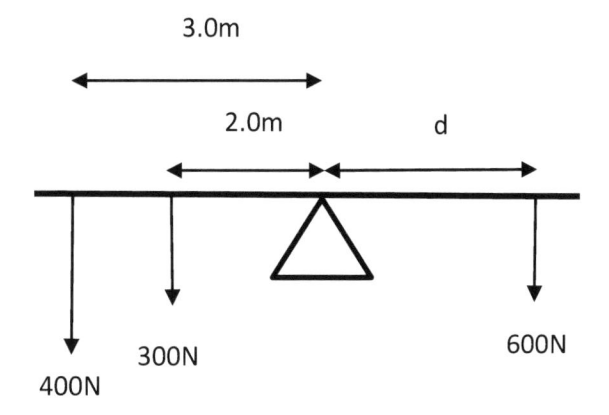

```
           3.0m
      ←——————————→
           2.0m       d
      ←————————→←————————→
  ━━━━━━━━━━━━━━━━━━━━━━━━
               △
   ↓    ↓              ↓
       300N           600N
  400N
```

Figure 6 shows a see-saw in equilibrium. Calculate the distance, **d**.

4) A see-saw has 3 people on it. To the left of the pivot is Bill who has a mass of 50kg (3m from the pivot) and Ben of mass 60kg (2m from the pivot). To the right of the pivot is Brian who has a mass of 90kg.

Calculate how far Brian is from the pivot if the see-saw is balanced. (Hint: don't forget that M = F x d, where F is a force). (g_{earth} = 10N/kg)

5)
Figure 7

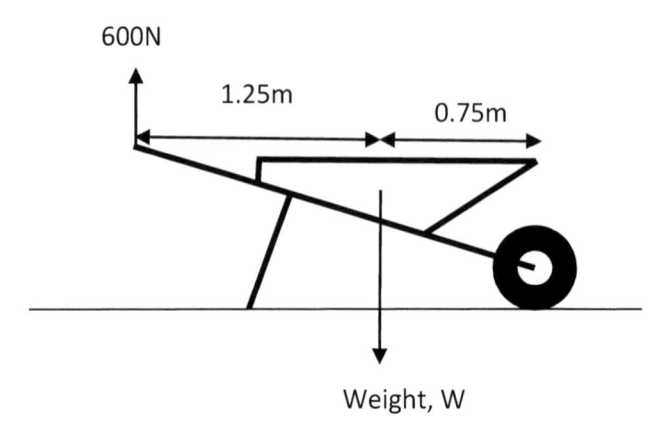

600N

1.25m

0.75m

Weight, W

Figure 7 shows a diagram of a wheelbarrow. If a force of 600N is needed to just lift the wheelbarrow off the ground, calculate the weight of the wheelbarrow and contents. If the wheelbarrow has a mass of 20kg, what is the mass of the contents? (g_{earth} = 10N/kg)

20. F = k x e

Hooke's law relates to the elasticity of materials. It states that the applied force (to a spring or wire for example) is directly proportional to extension as long as the limit of proportionality is not exceeded. If the limit of proportionality is exceeded, the applied force is no longer directly proportional to extension.

Note:
If a relationship between quantities is 'directly proportional' it means, for example, that:
if the force doubles the extension doubles
if the force trebles, the extension trebles
if the force halves, the extension halves

'Force is directly proportional to extension' can be written as **F ∝ e**. By introducing a constant, **k**, we can now write, **F = k x e**. 'k' can be known as the 'spring constant', 'stiffness constant' or the 'force constant'. I will refer to it as the 'spring constant'.

F is the applied force measured in newtons, N.
k is the 'spring constant' measured in newtons per metre, N/m (or Nm^{-1}).
e is the extension measured in metres, m.

Let's take a look at rearranging the formula:

$$F = k \times e$$

To make **k** the subject of the formula, we need to divide both sides by **e**. This gives:

Physics calculations for GCSE & IGCSE

$$\frac{F}{e} = k$$

Therefore,

$$k = \frac{F}{e}$$

Going back to the original formula, F = k x e, if we want to make **e** the subject of the formula, we need to divide both sides by **k**. This gives:

$$\frac{F}{k} = e$$

Therefore,

$$e = \frac{F}{k}$$

Example 1

A spring has a spring constant of 10N/m and when a force is applied to the spring, it extends by 5cm. Calculate the force applied.

k = 10N/m
e = 5.0cm = 0.05m (don't forget that extension must be in metres,m)

To convert cm to m you divide by 100.

F = ?

$$F = ke$$

$$F = 10 \times 0.05$$
$$F = 0.5N$$

Example 2

A spring has an original length of 5cm. When a force of 6N is applied to the spring, the spring extends to a length of 25cm. Calculate the spring constant.

Note:
To find the extension of the spring we need to use the following formula:

Extension = extended length – original length

e = 25 – 5 = 20cm

e = 20cm = 0.20m (remember, the extension must be in metres, m)
F = 6N
k = ?

$$F = ke$$

$$k = \frac{F}{e}$$

$$k = \frac{6}{0.20}$$

$$k = 30\text{N/m}$$

Example 3

A spring has a spring constant of 50N/m. If a force of 20N is applied to it, calculate the extension.

k = 50N/m
F = 20N
e = ?

$$F = ke$$

$$e = \frac{F}{k}$$

$$e = \frac{20}{50}$$

$$e = 0.40\text{m}$$

Example 4

A spring extends 3.0cm when a force of 0.45N is applied to it.
    a)   Calculate the spring constant.
    b)   If a force of 1.35N is applied, calculate the new extension.

    a)   e = 3.0cm = 0.030m
F = 0.45N
k = ?

$$F = ke$$

$$k = \frac{F}{e}$$

$$k = \frac{0.45}{0.03}$$

$$k = 15\text{N/m}$$

    b)   k = 15N/m   (our answer from the first part of the question)
          F = 1.35N
          e = ?

$$F = ke$$

$$e = \frac{F}{k}$$

$$e = \frac{1.35}{15}$$

$$e = 0.090\text{m}$$

Questions

1) A spring has a spring constant of 12N/m and when a force is applied to the spring, it extends by 6.0cm. Calculate the force applied.
2) A spring has an original length of 10cm. When a force of 20N is applied to the spring, the length is now 50cm. Calculate the spring constant.
3) A spring has a spring constant of 100N/m. Calculate the extension if the force applied is:
   a) 20N b) 30N c) 40N d) 50N e) 25N
4) A spring extends 5.0cm when a force of 2.0N is applied to it.
a) Calculate the spring constant.
b) If a force of 4.0N is applied, calculate the new extension.
5) A spring extends 10.0cm when a force of 5.0N is applied to it.
a) Calculate the spring constant.
b) If a force of 25.0N is applied, calculate the new extension.
6) A spring has a spring constant of 40N/m. Calculate the force applied to the spring if the extension is:
   a) 0.10m b) 200mm c) 30cm d) 0.50m e) 800mm

21. Centripetal Force

When an object is moving in a circle, the force acting on it is directed towards the centre of the circle. This force is called the centripetal force. The formula for the centripetal force is given below:

$$\text{Centripetal force} = \frac{\text{mass} \times \text{speed}^2}{\text{radius}}$$

Centripetal force is measured in newtons, N.
Mass is measured in kilograms, kg.
Speed is measured in metres per second, m/s.
Radius is measured in metres, m.

This formula can be written as:

$$F = \frac{mv^2}{r}$$

Let's take a look at rearranging the formula. If we want to make **r** the subject of the formula, we need to multiply both sides by **r**. This gives:

$$Fr = \frac{mv^2 r}{r}$$

The **r**'s on the right hand side of the formula will cancel, to give:

$$Fr = mv^2$$

Due to the 'F and r' being multiplied together, to get rid of the **F** we need to divide **both** sides by **F**. This gives:

$$\frac{Fr}{F} = \frac{mv^2}{F}$$

The **F**'s on the left hand side of the formula cancel. This gives:

$$r = \frac{mv^2}{F}$$

If we now look at making **m** the subject of the formula. To avoid repeating steps I will go to the step earlier where we had:

$$Fr = mv^2$$

To get the **m** by itself, due to the 'm and v²' being multiplied together we need to divide both sides by **v²**. This gives:

$$\frac{Fr}{v^2} = \frac{mv^2}{v^2}$$

The **v²**'s on the right hand side of the formula cancel. This will then give:

$$\frac{Fr}{v^2} = m$$

Therefore,

$$m = \frac{Fr}{v^2}$$

If we now take a look at rearranging the formula for **v²**. Again, to avoid repeating steps I will go to the step above where we had:

$$Fr = mv^2$$

To get the **v²** by itself, due to the 'm and v²' being multiplied together we need to divide both sides by **m**. This gives:

$$\frac{Fr}{m} = \frac{mv^2}{m}$$

The **m**'s on the right hand side of the formula cancel. This gives:

$$\frac{Fr}{m} = v^2$$

Therefore,

$$v^2 = \frac{Fr}{m}$$

In order to find **v**, we need to square root both sides. This will now give:

$$v = \sqrt{\frac{Fr}{m}}$$

Example 1

A conker of mass 50g attached to a piece of string is whirled in a circle of radius 30cm, at a speed of 4.0m/s. Calculate the centripetal force.

m = 50g = 0.050kg **(to convert g to kg you divide by 1000)**

r = 30cm = 0.30m   (to convert cm to m you divide by 100)
v = 4.0m/s
F = ?

$$F = \frac{mv^2}{r}$$

$$F = \frac{0.050 \times 4.0^2}{0.30}$$

$$F = \frac{0.050 \times 16}{0.30}$$

$$F = 2.67N$$

$$F = 2.7N$$

(2 significant figures are the minimum number given in the question, so the answer is given to 2 significant figures)

Example 2

In throwing the hammer of mass 7.3kg an athlete whirls a steel ball in a circle of radius 2.0m. If the centripetal force on the ball is 3285N, calculate the speed of the ball.

m = 7.3kg
r = 2.0m
F = 3285N
v = ?

$$F = \frac{mv^2}{r}$$

$$v^2 = \frac{Fr}{m}$$

$$v^2 = \frac{3285 \times 2.0}{7.3}$$

$$v^2 = 900$$

If we now square root both sides, we get:

$$v = 30 m/s$$

My advice, to avoid any silly calculator mistakes, is to calculate v^2 first. Once you have an answer for v^2, you just need to square root that answer as above.

Example 3

A 200m runner is running the bend in the first lane of a 200m race. The radius of curvature is 36.5m. If the speed of the athlete is 10m/s and the centripetal force acting on them is 245N, calculate the athletes mass.

r = 36.5m
v = 10 m/s
F = 245N
m = ?

$$F = \frac{mv^2}{r}$$

$$m = \frac{Fr}{v^2}$$

$$m = \frac{245 \times 36.5}{10^2}$$

$$m = \frac{8942.5}{100}$$

m = 89.425kg

m = 89kg

Example 4

The International Space Station (ISS) has a mass of 419,455kg. If the orbital speed is 7.71km/s and the centripetal force acting on it is 3.69MN, calculate the radius of the orbit from the centre of the earth. If the earth has a radius of 6.40×10^6m, calculate the height of the orbit above sea level.

m = 419,455kg
v = 7.71km/s = 7710 m/s ('kilo' means **x10³** (or **x 1000**))
F = 3.69MN = 3.69 x 10⁶N
r = ?

$$F = \frac{mv^2}{r}$$

$$r = \frac{mv^2}{F}$$

$$r = \frac{419,455 \times 7710^2}{3.69 \times 10^6}$$

$$r = \frac{419,455 \times 59444100}{3.69 \times 10^6}$$

r = 6757215.438

r = 6.76 x 10⁶m

To calculate the height above sea level, we need to subtract the radius of the earth from this value.

Height above sea level = 6.76 x 10⁶ - 6.40 x 10⁶

Height above sea level = 6,760,000 – 6,400,000

Height above sea level = 360,000m

Questions

1) An object which has a mass of 2.0kg is being rotated in a circle at a speed of 3.0m/s. Calculate the centripetal force, if the radius is:
   a) 9.0cm b) 0.36m c) 18cm d) 720mm e) 0.45m
2) An object is being whirled in a circle of radius 0.50m with a speed of 4.0 m/s. Calculate the mass of the object if the centripetal force is:
   a) 5.0N b) 15.0N c) 50.0N d) 75.0N e) 100.0N
3) An object of mass 5.0kg has a centripetal force acting on it of 100N. Calculate the radius of the circle if the speed of the object is:
   a) 2.0 m/s b) 3.0 m/s c) 4.0 m/s d) 5.0 m/s e) 10 m/s
4) In throwing the hammer of mass 7.3kg an athlete whirls a steel ball in a circle or radius 2.0m. If the centripetal force on the ball is 2280N, calculate the speed of the ball.
5) A 200m runner is running the bend in the first lane of a 200m race. The radius of curvature is 36.5m. If their speed is 9.0m/s and the centripetal force acting on them is 178N, calculate the athletes mass.
6) The mean distance from the centre of the earth to the centre of the moon is 3.844×10^8m. If the moon has a mass of 7.4×10^{22}kg and the centripetal force acting on the moon is 2.01×10^{20}N, calculate the orbital speed of the moon.
7) A Geostationary satellite of mass 3238kg has an orbital speed of 3.07km/s. If the centripetal force acting on it is 720N, calculate the radius of the orbit (remember that this is from the centre of the earth). If the radius of the earth is 6.4×10^6m, calculate the orbit height from the earth's surface.

22. Density

The formula for density is given below:

$$density = \frac{mass}{volume}$$

Density is measured in kilograms per metre cubed, kg/m^3 (or kgm^{-3}).
Mass is measured in kilograms, kg.
Volume is measured in metres cubed, m^3.

Note:
Sometimes density can be given in grams per centimetre cubed, g/cm^3. If this is the case, mass would be measured in grams, g, and volume would be measured in centimetre cubed, cm^3. It would be made very clear in the question if these units were the ones that were required.

It is also possible that you will be given the mass in grams and the volume in cm^3 or mm^3 and the density will need to be in kg/m^3. In this situation you need to be able to convert units.

To convert grams, g to kilograms, kg, you divide by 1000.

Converting cm^3 to m^3.

1cm = 0.01m (to convert cm to m you divide by 100)

Therefore,

$1cm^3 = 1cm \times 1cm \times 1cm$, thus,

$1cm^3 = 0.01 \times 0.01 \times 0.01 = 1 \times 10^{-6} m^3$

So, if you are given a volume in cm³ and you want to convert to m³, you simply multiply by **1 x 10⁻⁶** (or **0.000001**). Therefore,

$20cm^3 = 20 \times 1 \times 10^{-6} = 2 \times 10^{-5} m^3$ (0.00002m³)
$50cm^3 = 50 \times 1 \times 10^{-6} = 5 \times 10^{-5} m^3$ (0.00005m³)
$1000cm^3 = 1000 \times 1 \times 10^{-6} = 1 \times 10^{-3} m^3$ (0.001m³)

Converting mm³ to m³.

$1mm = 0.001m$ (to convert mm to m you multiply by 0.001 or divide by 1000)

Therefore,

$1mm^3 = 1mm \times 1mm \times 1mm$

$1mm^3 = 0.001 \times 0.001 \times 0.001 = 1 \times 10^{-9} m^3$

So, if you are given a volume in mm³ and you want to convert to m³, you simply multiply by **1 x 10⁻⁹** (or **0.000000001**). Therefore,

$50mm^3 = 50 \times 1 \times 10^{-9} = 5 \times 10^{-8} m^3$ (0.00000005m³)
$200mm^3 = 200 \times 1 \times 10^{-9} = 2 \times 10^{-7} m^3$ (0.0000002m³)
$1000mm^3 = 1000 \times 1 \times 10^{-9} = 1 \times 10^{-6} m^3$ (0.000001m³)

Going back to the formula,

$$density = \frac{mass}{volume}$$

This can be written as:

$$\rho = \frac{m}{V}$$

Where ρ, (Greek letter 'rho') is the symbol used for density.
Taking a look at rearranging the formula, if we want to make **m** the subject of the formula, we need to multiply both sides by **V**. This gives:

$$\rho \times V = m$$

Therefore,

$$m = \rho V$$

If we want to make **V** the subject of the formula, from the formula above, $m = \rho V$, we need to divide both sides by ρ. This gives:

$$\frac{m}{\rho} = V$$

Therefore,

$$V = \frac{m}{\rho}$$

Example 1

A block of platinum has a mass of 0.2145kg and a volume of 10.00cm³. Calculate the density of platinum (in kg/m³)

m = 0.2145kg
V = 10.00cm³ = 10.00 x 1 x 10⁻⁶ = 1 x 10⁻⁵m³ (0.00001m³)
ρ = ?

$$\rho = \frac{m}{V}$$

$$\rho = \frac{0.2145}{0.00001}$$

ρ = 2.145 x 10⁴ kg/m³

ρ = 21,450 kg/m³

Example 2

Gold has a density of 19.3 x 10³ kg/m³. If an object has a volume of 5.00 x 10⁻⁸m³, calculate its mass.

ρ = 19.3 x 10³ kg/m³ = 19,300kg/m³
V = 5.00 x 10⁻⁸m³ (0.00000005m³)
m = ?

$$\rho = \frac{m}{V}$$

$$m = \rho V$$

$$m = 19{,}300 \times 0.00000005$$

m = 9.65 x 10⁻⁴kg

Or,

m = 0.000965kg

Example 3

Pure iron has a density of 7.87g/cm³. Calculate the volume of iron (in cm³), if the mass is 200g.

ρ = 7.87g/cm³
m = 200g
V = ?

$$\rho = \frac{m}{V}$$

$$V = \frac{m}{\rho}$$

$$V = \frac{200}{7.87}$$

$$V = 25.4 \text{cm}^3$$

Questions

1) Aluminium has a density of 2710kg/m³. Calculate the mass if the volume is:
   a) 50.0cm³ b) 2.00m³ c) 10.00m³ d) 20.0cm³ e) 1000mm³
2) Copper has a density of 8930kg/m³. Calculate the volume of copper in m³ if the mass is:
   a) 20.0g b) 50.0kg c) 1000kg d) 60 x 10⁻³kg e) 4.0kg
3) A diamond has a mass of 3.30 x 10⁻³kg and a volume of 1cm³. Calculate the density of diamond in kg/m³.
4) A titanium bike frame has a mass of 2.0kg. If the density of titanium is 4,540kg/m³, calculate the volume of titanium.
5) A solid silver trophy has a mass of 0.315kg and a volume of 30.0cm³. Calculate the density of silver in kg/m³.
6) Mercury has a density of 13,546kg/m³. Calculate the mass of mercury if the volume is 50.00 x 10⁻⁶m³.
7) Oxygen has a density of 1.429kg/m³. Calculate the volume of 1000g of oxygen.

23. Pressure

The pressure acting on a surface is defined as the force acting per unit area. The formula for this is given below:

$$pressure = \frac{force}{area}$$

Pressure is measured in pascals, Pa (note also that an alternative unit is the newton per metre squared, N/m² (or Nm⁻²)). 1Pa = 1N/m².
Force is measured in newtons, N.
Area is measured in metres squared, m².

This can be written as:

$$P = \frac{F}{A}$$

Taking a look at rearranging the formula, if we want to make **F** the subject of the formula, we need to multiply both sides by **A** (this is because the **F** is being divided by the **A**, and we have to do the 'inverse' or 'opposite', which is multiply). This gives:

$$P \times A = F$$

Therefore,
$$F = P \times A$$

Using the above formula, $F = P \times A$, to make **A** the subject of the formula, we need to divide both sides by **P**. This gives:

$$\frac{F}{P} = A$$

Therefore,
$$A = \frac{F}{P}$$

Converting areas

Converting cm² to m²

1cm = 0.01m (to convert cm to m you divide by 100)

1cm² = 1cm x 1cm, therefore,

1cm² = 0.01 x 0.01 = 1 x 10⁻⁴m² (0.0001m²)

Therefore to convert cm² to m², you multiply by **1 x 10⁻⁴** (or **0.0001**). A few examples are given below:

20cm² = 20 x 1 x 10⁻⁴ = 2 x 10⁻³m² (0.002m²)
150cm² = 150 x 1 x 10⁻⁴ = 0.015m²
1000cm² = 1000 x 1 x 10⁻⁴ = 0.1m²

Converting mm² to m²

1mm = 0.001m (to convert mm to m you multiply by 0.001 or divide by 1000)

1mm² = 1mm x 1mm, therefore,

1mm² = 0.001 x 0.001 = 1 x 10⁻⁶m² (0.000001m²)

Therefore to convert mm² to m², you multiply by **1 x 10⁻⁶** (or **0.000001**). A few examples are given below:

30mm² = 30 x 1 x 10⁻⁶ = 3 x 10⁻⁵m² (0.00003m²)
200mm² = 200 x 1 x 10⁻⁶ = 2 x 10⁻⁴m² (0.0002m²)
1000mm² = 1000 x 1 x 10⁻⁶ = 1 x 10⁻³m² (0.001m²)

Example 1

A man of mass 100kg is standing on the floor with flat shoes which have a total area of 0.05m². Calculate the pressure exerted on the floor. (g_{earth} = 10N/kg)

The first thing we need to do is calculate the weight of the man. Don't forget that weight is a force!

From section 9,

$$W = mg$$

$$W = 100 \ x \ 10$$

$$W = 1000N$$

F = W = 1000N
A = 0.05m²
P = ?

$$P = \frac{F}{A}$$

$$P = \frac{1000}{0.05}$$

$$P = 2 \times 10^4 \text{ Pa}$$

$$P = 20{,}000 \text{Pa}$$

Example 2

A container of area 15m² exerts a pressure on the ground of 2kPa. Calculate the weight of the container.

A = 15m²
P = 2kPa = 2 x 10³ Pa = 2000Pa
F = W = ?

$$P = \frac{F}{A}$$

$$F = P \times A$$

$$F = 2000 \times 15$$

$$F = 30{,}000 \text{N}$$

Therefore the weight of the container is 30,000N.

Example 3

An elephant has a weight of 56.0kN. If the total area of the elephants feet is 1400cm², calculate the pressure that the elephant exerts on the ground.

F = W = 56.0kN = 56.0 x 10³N = 56,000N
A = 1400cm² = 1400 x 1 x 10⁻⁴ = 0.14m²
P = ?

$$P = \frac{F}{A}$$

$$P = \frac{56{,}000}{0.14}$$

$$P = 400{,}000 \text{Pa}$$

Example 4

A tractor has a weight of 16,000N. If the pressure exerted on the ground is 1.20 x 10⁴Pa, calculate the contact area.

F = W = 16,000N
P = 1.20 X 10⁴Pa = 12,000Pa
A = ?

$$P = \frac{F}{A}$$

$$A = \frac{F}{P}$$

$$A = \frac{16,000}{12,000}$$

$$A = 1.33 m^2$$

Questions (g_{earth} = 10N/kg)

1) A box has a weight of 100N. Calculate the pressure if the area of contact is:
   a) 20cm² b) 50cm² c) 0.01m² d) 15,000mm² e) 0.02m²
2) An object exerts a pressure on the ground of 40kPa. Calculate the weight of the object, if the area in contact with the ground is:
   a) 10cm² b) 20,000mm² c) 0.025m² d) 15cm² e) 0.05m²
3) A box has a mass of 20kg. Calculate the area in contact with the ground, if the pressure is:
   a) 10,000Pa b) 1.5kPa c) 4.0 x 10³Pa d) 500Pa e) 200Pa
4) A skier has a mass of 80kg. If the pressure exerted on the ground by the ski's is 3.2kPa, calculate the area in contact with the ground.
5) A woman of weight 500N is wearing stiletto heals and is balancing on both heals. If the area of 1 stiletto heal is 100mm², calculate the pressure exerted on the ground.
6) A garden chair has 4 legs, each with an area of 2cm² in contact with the ground. The chair has a mass of 5.0kg. If a woman of mass 55.0kg sits on the chair, calculate the pressure on the ground.
7) A slab has an area in contact with the ground of 810,000mm². If the pressure exerted on the ground is 370.4Pa, calculate the weight of the slab.

24. Pressure variation with depth in fluids.

A fluid is a substance that can flow. Both liquids and gases are classed as fluids. The pressure in a fluid increases with depth and all points at the same depth in the fluid will have the same pressure.
The formula for the pressure variation with depth is given by:

$$pressure = height \; x \; density \; x \; gravitational \; field \; strength$$

Pressure is measured in pascals, Pa (note also that an alternative unit is the newton per metre squared, N/m² (or Nm⁻²)). 1Pa = 1N/m².
Height is measured in metres, m. (this is the vertical height)
Gravitational field strength is measured in newtons per kilogram, N/kg.

This can be written as:

$$p = h \; x \; \rho \; x \; g$$

Where ρ, (Greek letter 'rho') is the symbol used for density.

Note:
The above equation is **not valid for gases** when the **height is large**. This is because in the above equation the density must be a constant value. For example, as you go higher above the earth's surface, the density of the air gets

lower and therefore does not have a constant value as the height increases. The above formula uses density values that must be constant.

From the above equation, we can also use:

$$\Delta p = h \times \rho \times g$$

Where Δp, is the change in pressure between 2 points separated by a vertical height, **h**. (Δ, is the Greek letter 'delta' which means 'change in').

Taking a look a rearranging the formula, if we want to make **h** the subject of the formula, we need to divide both sides by '$\rho \times g$' (**h** is multiplied by '$\rho \times g$', so we need to divide both sides by '$\rho \times g$', because we have to do the 'inverse' or 'opposite'). This will give:

$$\frac{\Delta p}{\rho g} = h$$

Therefore,

$$h = \frac{\Delta p}{\rho g}$$

If we want to make **ρ** the subject of the formula, going back to the original formula, $\Delta p = h \times \rho \times g$, we need to divide both sides by '$h \times g$'. This gives:

$$\frac{\Delta p}{hg} = \rho$$

Therefore,

$$\rho = \frac{\Delta p}{hg}$$

If we want to make **g** the subject of the formula, going back to the original formula, $\Delta p = h \times \rho \times g$, we need to divide both sides by '$h \times \rho$'. This gives:

$$\frac{\Delta p}{h\rho} = g$$

Therefore,

$$g = \frac{\Delta p}{h\rho}$$

Example 1

Water has a density of 1000kg/m³. If a diver goes 10m below the surface of the water, calculate the pressure difference. (g_{earth} = 10N/kg)

ρ = 1000kg/m³
h = 10m
g = 10N/kg
Δp = ?

$$\Delta p = h \times \rho \times g$$

$$\Delta p = 10 \times 1000 \times 10$$

$$\Delta p = 100,000 Pa$$

Example 2

A diver in sea water, dives from a depth of 10.00m to 30.00m below the surface of the water. If the pressure difference in doing this is 205.0kPa, calculate the density of sea water. (g_{earth} = 10.00N/kg)

h = 30.00 − 10.00 = 20.00m (remember that **h** is the difference in height)
Δp = 205.0kPa = 205.0 x 10³Pa = 205,000Pa
g = 10.00N/kg
ρ = ?

$$\Delta p = h \times \rho \times g$$

$$\rho = \frac{\Delta p}{hg}$$

$$\rho = \frac{205,000}{20.0 \times 10.00}$$

ρ = 1025kg/m³

Example 3

Liquid mercury has a density of 13,546kg/m³. An object is placed in the mercury, so that the pressure difference is 6773Pa. Calculate the difference in height. (g_{earth} = 10N/kg)

ρ = 13,546kg/m³
Δp = 6773Pa
g = 10N/kg
h = ?

$$\Delta p = h \times \rho \times g$$

$$h = \frac{\Delta p}{\rho g}$$

$$h = \frac{6773}{13,546 \times 10}$$

h = 0.050m

Example 4

Paraffin oil has a density of 800kg/m³. When an object is placed 200mm below the surface of the oil, the pressure difference is 1.600kPa. Calculate the gravitational field strength.

ρ = 800kg/m³
h = 200mm = 200 x 10⁻³m = 0.200m
Δp = 1.600kPa = 1.600 x 10³Pa = 1600Pa
g = ?

$$\Delta p = h \times \rho \times g$$

$$g = \frac{\Delta p}{h\rho}$$

$$g = \frac{1600}{0.200 \times 800}$$

$$g = 10.0 \text{N/kg}$$

Questions (g_earth = 10N/kg)

1) Water has a density of 1000kg/m³. Calculate the pressure difference if the change in height is:
    a) 5.0m b) 20.0m c) 50.0m d) 40cm e) 500mm
2) Olive oil has a density of 920kg/m³. Calculate the change in height if the pressure difference is:
    a) 18,400Pa b) 4600Pa c) 2.3kPa d) 1.840kPa e) 920Pa
3) Sea water has a density of 1025kg/m³. When a submarine dives under water, the pressure difference is 20.50MPa. Calculate the depth of the submarine.
4) An object is placed in liquid mercury at a depth of 0.400m. Calculate the density of mercury, if the pressure difference is 54.184kPa.
5) Turpentine or 'Turps' is a liquid at room temperature of density 870kg/m³. If an object is placed 100mm below the surface of the liquid, calculate the change in pressure.

25. Hydraulics

If you have not already done so, it would be worth reading section 23 before looking at this section.

A hydraulic system depends on 2 principles:
1) Liquids are incompressible (can't be squashed).
2) At any point in a liquid, the same pressure acts equally in all directions.

Hydraulic systems are very useful as they can multiply forces (act as force multipliers), for example, in hydraulic car brakes and car jacks. An example of a hydraulic jack is given below:

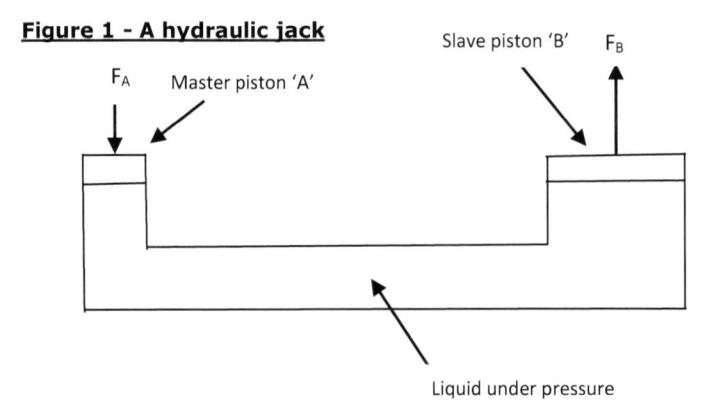

Figure 1 - A hydraulic jack

Slave piston 'B'   F_B
F_A   Master piston 'A'

Liquid under pressure

In figure 1 above, a downward force is applied to the master piston 'A'. This puts the liquid under pressure and this pressure is transmitted through the liquid. The pressure at 'A' is therefore the same pressure that acts on the slave piston 'B'. Due to the fact that the area of piston 'B' is greater than the area of piston 'A', the force on piston 'B' is greater. This system therefore acts as a force multiplier and a small force applied at piston 'A' can be used to apply a large force at 'B', making it much easier for you to lift an object.

Example 1

Figure 1 shows a diagram of a hydraulic jack. If a force of 100N is applied to master piston 'A' which has a cross sectional area (referred to as just 'area' from now on in this section) of 50cm², calculate the force exerted on slave piston 'B' if the area of B is 200cm².

The first thing we need to do is to calculate the pressure at 'A'. (Note that on these questions don't bother to convert the area into m². If the area is given in cm², calculate the pressure in N/cm² (newtons per centimetre squared). If the area is given in m², calculate the pressure in N/m² (newtons per metre squared)

Pressure at 'A'

From section 23, the formula for pressure is given by:

$$P = \frac{F}{A}$$

F = 100N
A = 50cm²
P = ?

$$P = \frac{100}{50}$$

P = 2N/cm²

This pressure gets transmitted through the liquid, therefore the pressure acting on slave piston 'B' is the same as what we have just calculated at 'A'.

Therefore, **pressure at 'B' = 2N/cm²**
Now we have to calculate the force at 'B'

Force at 'B'

P = 2N/cm²
A = 200cm²
F =

$$P = \frac{F}{A}$$

$$F = P \, x \, A$$

$$F = 2 \, x \, 200$$

F = 400N

So, a force of 100N was applied to master piston 'A' and a force of 400N is applied to slave piston 'B'. The force has been multiplied! This is obviously a very useful machine, as in the case of the above example, it can be used to make it easier to lift objects.

A quick way to work out the force is to compare the areas of the pistons. Area 'B' is 4 times greater than area 'A' (200/50 = 4). Another way of saying this is that the ratio of area 'B' to area 'A' is 4:1 (4 to 1). The ratio of the forces will be exactly the same as the ratio of the areas, 4:1. This means that the force at 'B' will be 4 times greater than at 'A'. So, if the force on piston 'A' is 100N, the force on piston 'B' will be 400N.

Alternatively, there is another method. If the pressure at 'A' equals the pressure at 'B', we can write:

$$P_A = P_B$$

But,

$$P = \frac{F}{A}$$

Therefore, we can also write:

$$\frac{Force_A}{Area_A} = \frac{Force_B}{Area_B}$$

$$\frac{F_A}{A_A} = \frac{F_B}{A_B}$$

This is another useful formula, although you may find that the other methods are easier depending on your mathematical skills. The above formula,

$$\frac{F_A}{A_A} = \frac{F_B}{A_B}$$

can be rearranged to find F_A, A_A, F_B or A_B. As I have given 2 methods for solving hydraulics problems already, I am not going to go through how to rearrange this formula now. I will explain how to rearrange this type of formula when we look at the 'transformer' formula in section 40. If you feel confident in your ability to use this formula then please try it out. If you are not confident, but would like to attempt to use it, go to section 40.

Questions

1) A hydraulic jack has a master piston of area 20cm², with a force acting on it of 50N. Calculate the force acting on the slave piston, if the area of the slave piston is:
   a) 40cm² b) 80cm² c) 100cm² d) 200cm² e) 1000cm²
2) A hydraulic jack has a master piston of area 0.50m², with a force acting on it of 200N. Calculate the force acting on the slave piston, if the area of the slave piston is:
   a) 1.0m² b) 2.0m² c) 1.5m² d) 2.5m² e) 3.0m²
3) A hydraulic jack has a master piston of area 100cm², with a force acting on it of 500N. If the force acting on the slave piston is 2000N, calculate the area of the slave piston.
4) A hydraulic braking system has a master piston of area 4.00cm². The force applied to this piston is 500N.

a) What is the pressure at the master piston
b) What is the pressure at the slave piston
c) What is the force applied to the slave piston, if the area of the slave piston is 16.00cm².
5) A hydraulic braking system has a master piston which has an area of 5.0cm². If the slave piston has an area of 20.0cm² and the force acting on it is 600N, calculate the force on the master piston.
6) A hydraulic jack has a master piston of area 100cm², with a force acting on it of 1000N. If the force acting on the slave piston is 6000N, calculate the area of the slave piston.
7) A hydraulic braking system has a master piston of area 3.0cm². The force applied to this piston is 600N.
a) What is the pressure at the master piston
b) What is the pressure at the slave piston
c) What is the force applied to the slave piston, if the area of the slave piston is 18cm².

26. Boyle's law

Boyle's law states the following:
For a fixed mass of gas at constant temperature, the product of pressure and volume is constant.
This can be written as:

$$pV = \text{constant}$$

If we let the constant equal **k**, we can write this as:

$$pV = k$$

Dividing both sides by V, we get:

$$p = \frac{k}{V}$$

Mathematically, we can now say that pressure is inversely proportional to volume,

$$p \propto \frac{1}{V}$$

What this means, is that, for example, if we double the volume, the pressure will halve. If we triple the volume, the pressure will be one third.

If we go back to **pV = k**, an alternative way of writing this is:

$$p_1V_1 = p_2V_2$$

This shows that the product (multiplication) of pressure and volume is constant. p_1 multiplied by V_1 is always equal to p_2 multiplied by V_2.
In this formula, p_1 and V_1 are the initial pressure and volume of the system, while p_2 and V_2 are the final pressure and volume of the system. With this formula, pressure can be measured in any unit of pressure, so long as the same unit is used for both p_1 and p_2. Examples of units used for pressure are: Pa (pascals), N/m^2, N/cm^2, atmospheres, bar, mmHg (millimetres of mercury) and psi (pounds per square inch).

Similarly, volume can be measured in any unit of volume, so long as the same unit is used for both V₁ and V₂. Examples of units used for volume are: m³, cm³, ml (millilitres) and l (litres).

We will be using this formula (p₁V₁ = p₂V₂) to help us solve problems on Boyle's law. Taking a look at rearranging the formula, if we want to make **p₁** the subject of the formula, dividing both sides by **V₁** gives:

$$p_1 = \frac{p_2 V_2}{V_1}$$

Going back to the original formula, p₁V₁ = p₂V₂, if we want to make **V₁** the subject of the formula, dividing both sides by **p₁** gives:

$$V_1 = \frac{p_2 V_2}{p_1}$$

Similarly, making **p₂** the subject of the formula gives:

$$p_2 = \frac{p_1 V_1}{V_2}$$

While making **V₂** the subject of the formula will give:

$$V_2 = \frac{p_1 V_1}{p_2}$$

Example 1

A gas has a volume of 4m³ at a pressure of 3 atmospheres. If the pressure of the gas is increased to 6 atmospheres and the mass and temperature of the gas remain constant, what volume would the gas occupy?

I will show 2 methods for solving this problem.

Method 1

We have just shown above that:

$$p \propto \frac{1}{V}$$

It follows also from this that:

$$V \propto \frac{1}{p}$$

This has come from the fact that, from earlier:

$$pV = k$$

Dividing both sides by **p** gives:

$$V = \frac{k}{p}$$

From this formula, we can say that volume is inversely proportional to pressure or,

$$V \propto \frac{1}{p}$$

From this relationship, if the pressure has doubled, the volume will halve, so the volume will be 2m³.

Method 2

p_1 = 3 atmospheres
V_1 = 4m³
p_2 = 6 atmospheres
V_2 = ?

$$p_1 V_1 = p_2 V_2$$

$$V_2 = \frac{p_1 V_1}{p_2}$$

$$V_2 = \frac{3 \times 4}{6}$$

$$V_2 = \frac{12}{6}$$

$$V_2 = 2 m^3$$

Example 2

A gas is at a pressure p_1, and has a volume of 500cm³. The volume is then reduced to 400cm³ and the new pressure is 937.5mmHg. If the mass and temperature of the gas remain constant, calculate p_1.

p_1 = ?
V_1 = 500cm³
p_2 = 937.5mmHg
V_2 = 400cm³

$$p_1 V_1 = p_2 V_2$$

$$p_1 = \frac{p_2 V_2}{V_1}$$

$$p_1 = \frac{937.5 \times 400}{500}$$

$$p_1 = \frac{375{,}000}{500}$$

$$p_1 = 750 mmHg$$

Questions

For all questions assume that the temperature and mass of the gas remain constant.

1) A gas has a volume of 3.0m³ at a pressure of 2.0 atmospheres. What volume would the gas occupy, if the pressure was increased to:
   a) 4.0 atmospheres b) 3.0 atmospheres c) 6.0 atmospheres
   d) 5.0 atmospheres e) 8.0 atmospheres
2) A gas is at a pressure of 1000mmHg and has a volume of 600cm³. The volume is then increased to 700cm³. Calculate the pressure.
3) A gas has an initial volume of V_1 and a pressure of 4 atmospheres. If the final volume is 400cm³ and pressure 3 atmospheres, calculate V_1.
4) A gas has a volume of 200cm³ and pressure of 1.00×10^5Pa. If the pressure is reduced to 0.80×10^5Pa, calculate the new volume.
5) A gas has an initial volume of 100cm³ and is at a pressure of 2.00×10^5Pa. If the volume is increased to 150cm³, calculate the pressure.

27. Pressure law

Before introducing the pressure law I need to explain the kelvin scale of temperature. With the kelvin scale of temperature 0K (zero kelvin) is the lowest temperature that in theory could be reached. This is also called absolute zero (absolute zero = -273°C).
1°C (1 degree Celsius) = 1K (1 kelvin). To convert from degrees Celsius to kelvin we use the following formula:

Temperature in kelvin, T = 273 + Celsius temperature

Therefore:
0°C will equal '273 + 0' which equals 273K.
20°C will equal '273 + 20' which equals 293K.
100°C will equal '273 + 100' which equals 373K.
-273°C will equal '273 + -273' which equals '273 – 273' which equals 0K.

The pressure law states the following:
For a fixed mass of gas at constant volume, pressure is directly proportional to temperature measured in kelvins.

This can be written as:
$$\frac{p}{T} = constant$$

Where **T** is the temperature measured in **kelvins**.
If we let the constant equal **k**, we can write this as:
$$\frac{p}{T} = k$$

Multiplying both sides by **T** gives:
$$p = kT$$

From this we can now say mathematically that pressure is directly proportional to temperature (in kelvins),

$$p \propto T$$

What this means, is that, for example, if the pressure is doubled the temperature is doubled. If the pressure is halved, the temperature is halved. If the pressure is tripled, the temperature is tripled.

If we go back to the formula,

$$\frac{p}{T} = k$$

another way of writing this is:

$$\frac{p_1}{T_1} = \frac{p_2}{T_2}$$

In this formula p_1 and T_1 are the initial pressure and temperature and p_2 and T_2 are the final pressure and temperature. As with Boyle's law, pressure can be measured in any unit of pressure, so long as the same unit is used for both p_1 and p_2. **Both T_1 and T_2 must be measured in kelvins.**

We will be using this formula to help us solve problems on the pressure law. Taking a look at rearranging the formula, if we want to make **p_1** the subject of the formula, multiplying both sides by **T_1** gives:

$$p_1 = \frac{p_2 T_1}{T_2}$$

From this, if we multiply both sides by **T_2** we get:

$$p_1 T_2 = p_2 T_1$$

If we now want to make **T_2** the subject of the formula, we need to divide both sides by **p_1**. This gives:

$$T_2 = \frac{p_2 T_1}{p_1}$$

If we now go back to **$p_1 T_2 = p_2 T_1$**, to make **p_2** the subject of the formula, we need to divide both sides by **T_1**. This gives:

$$\frac{p_1 T_2}{T_1} = p_2$$

Therefore,

$$p_2 = \frac{p_1 T_2}{T_1}$$

Again, if we go back to **$p_1 T_2 = p_2 T_1$**, to make **T_1** the subject of the formula, dividing both sides by **p_2** gives:

$$\frac{p_1 T_2}{p_2} = T_1$$

Therefore,

$$T_1 = \frac{p_1 T_2}{p_2}$$

Example 1

A gas is at a pressure of 1.00 x 10^{5}Pa and temperature 0°C. If the temperature is increased to 273°C, calculate the pressure assuming that the volume and mass of gas remain constant.

Firstly, all temperatures need to be in kelvin. To convert **°C to K** we use the following formula:

Temperature in kelvin, T = 273 + Celsius temperature

Therefore,

0°C will equal '273 + 0' which equals 273K.
273°C will equal '273 + 273' which equals 546K.

I will now show 2 methods for solving this problem.

Method 1

p_1 = 1.00 x 10^{5}Pa
T_1 = 273K
T_2 = 546K
p_2 = ?

$$\frac{p_1}{T_1} = \frac{p_2}{T_2}$$

$$p_2 = \frac{p_1 T_2}{T_1}$$

$$p_2 = \frac{1.00 \times 10^5 \times 546}{273}$$

$$p_2 = \frac{5.46 \times 10^7}{273}$$

$$p_2 = 2.00 \times 10^5 \text{Pa}$$

Or,

$$p_2 = 200{,}000 \text{Pa}$$

Method 2

We have just shown above that:

$$p \propto T$$

If we take a look at the **temperatures in kelvin**, the final temperature is twice as big as the initial temperature (546/273 = 2). Therefore, because pressure is directly proportional to temperature, if the temperature has doubled, the pressure will double. Therefore the pressure will be:

$1.00 \times 10^5 \times 2 = 2.00 \times 10^5 Pa$.

Therefore,
$$p_2 = 2.00 \times 10^5 Pa$$

Example 2

A gas is in a sealed container at an initial pressure of $0.50 \times 10^5 Pa$ and is at a temperature of -173°C. Calculate the temperature, if the pressure changes to $1.50 \times 10^5 Pa$, assuming that the volume and mass of gas remain constant.

Again, the first thing that we need to do is to convert the temperature from degree Celsius to kelvin. To convert **°C to K** we use the following formula:

$$\text{Temperature in kelvin, } T = 273 + \text{Celsius temperature}$$

Therefore,

-173°C will equal '273 + -173' which equals '273 – 173' which equals 100K.

I will again show 2 methods for solving this problem.

Method 1

$p_1 = 0.50 \times 10^5 Pa$
$T_1 = 100K$
$T_2 = ?$
$p_2 = 1.50 \times 10^5 Pa$

$$\frac{p_1}{T_1} = \frac{p_2}{T_2}$$

$$T_2 = \frac{p_2 T_1}{p_1}$$

$$T_2 = \frac{1.50 \times 10^5 \times 100}{0.50 \times 10^5}$$

$$T_2 = \frac{1.50 \times 10^7}{0.50 \times 10^5}$$

$$T_2 = 300K$$

Method 2

We have just shown above that:

$$p \propto T$$

If we take a look at the pressures, the final pressure is three times as big as the initial pressure ($1.50 \times 10^5 / 0.50 \times 10^5 = 3$). Therefore, because pressure is directly proportional to temperature, if the pressure has tripled, the temperature will triple. Therefore the temperature will be 100 x 3 = 300K.
Therefore,
$$T_2 = 300K$$

Questions

For all questions assume that the volume and mass of gas remain constant.
Remember that all temperatures need to be in kelvin.

1) A gas is at an initial pressure of 1.0 atmosphere and a temperature of 100K. Calculate the pressure of the gas if the temperature changes to:
   a) 200K b) 250K c) 300K d) 500K e) 1000K
2) A gas is at an initial pressure of 2.00×10^5Pa and an initial temperature of 300K. Calculate the temperature of the gas if the pressure changes to:
   a) 2.20×10^5Pa b) 2.50×10^5Pa c) 3.00×10^5Pa d) 2.80×10^5Pa e) 3.20×10^5Pa
3) A gas has an initial pressure of 2.00 atmospheres. If the pressure changes to 4.00 atmospheres and the temperature changes to 327°C, calculate the initial temperature.
4) A gas is at an initial temperature of 20.0°C. If the temperature changes to 313.0°C and the pressure changes to 3.0×10^5Pa, calculate the initial pressure.
5) A gas has an initial pressure of 0.40×10^5Pa. If the pressure changes to 1.20×10^5Pa and the temperature changes to 477°C, calculate the initial temperature.

28. Specific heat capacity

The specific heat capacity of a substance is the amount of energy required to raise 1kg of the substance by 1°C. The formula for this is given below:

$$energy\ transferred = mass \times specific\ heat\ capacity \times temperature\ change$$

Energy transferred is measured in joules, J.
Mass is measured in kilograms, kg.
Specific heat capacity is measured in joules per kilogram per degree Celsius, J/kg/°C (an alternative way of expressing this unit is 'joule per kilogram degree Celsius', J/kg°C).
Temperature change is measured in degrees Celsius, °C.

If we go back to the formula, this can be written as:

$$E = m \times c \times \theta$$

where **c** is the specific heat capacity and **θ** is the temperature change.

Taking a look at rearranging the formula, if we want to make **m** the subject of the formula, we need to divide both sides by '**c x θ**' or '**cθ**'. Remember that the m, c and θ are all multiplied together, so to get the **m** by itself you must do the 'opposite' or 'inverse', which is divide. This will give:

$$\frac{E}{c \times \theta} = m$$

Therefore,

$$m = \frac{E}{c \times \theta}$$

If we want to make **c** the subject of the formula, going back to the original formula, $E = m \times c \times \theta$, we need to divide both sides by '**m × θ**'. This gives:

$$\frac{E}{m \times \theta} = c$$

Therefore,

$$c = \frac{E}{m \times \theta}$$

Finally, if we want to make **θ** the subject of the formula, going back to the original formula, $E = m \times c \times \theta$, we need to divide both sides by '**m × c**'. This will give:

$$\frac{E}{m \times c} = \theta$$

Therefore,

$$\theta = \frac{E}{m \times c}$$

Example 1

500g of water at 20.0°C is heated to 100.0°C. Calculate the energy required to do this, if the specific heat capacity of water is 4200 J/kg/°C.

It is important to remember that in the formula, **θ**, is the **temperature change!**

$$\text{temperature change}, \theta = \text{final temperature} - \text{initial temperature}$$

Therefore, θ = 100.0 − 20.0 = 80.0°C.

Also, mass must be in kilograms. To convert grams, **g** into kilograms, **kg** you need to divide by 1000. Therefore, **m** = 500/1000 = 0.500kg.

m = 0.500kg
c = 4200 J/kg/°C
θ = 80.0°C

$$E = m \times c \times \theta$$

$$E = 0.500 \times 4200 \times 80.0$$

$$E = 1.68 \times 10^5 J$$

Or,

$$E = 168{,}000 J$$

Example 2

Brass has a specific heat capacity of 370J/kg/°C. If 740J of heat energy is supplied to the brass and the temperature of the brass rises by 20°C, calculate the mass of brass.

E = 740J
c = 370J/kg/°C

θ = 20°C
m = ?

$$E = m \times c \times \theta$$

$$m = \frac{E}{c \times \theta}$$

$$m = \frac{740}{370 \times 20}$$

My advice here is to calculate '370 x 20' first. A common mistake is to do '740÷370x20'. **This is wrong!** If you do this, going back to **BIDMAS** (refer back to section 5 if you have forgotten), multiplication and division have the same priority and you therefore do the calculation from left to right. This will give '740÷370 = 2. The 2 is then multiplied by 20 which gives 40! **This is incorrect!** By calculating '370 x 20' first, you will avoid this mistake. 370 x 20 = 7400. Therefore we get:

$$m = \frac{740}{7400}$$

m = 0.10kg

This is the correct answer and is very different from 40. When you get a calculation like below:

$$m = \frac{740}{370 \times 20}$$

the 740 is divided by '370 x 20'. If it helps, put the '370 x 20' in brackets, as using BIDMAS, you will calculate what is in the brackets first, as below:

$$m = \frac{740}{(370 \times 20)}$$

By calculating the brackets first, you will avoid the silly mistake that is mentioned above.

Example 3

An aluminium block of mass 5.0kg of specific heat capacity 900J/kg/°C is at 70°C. As it cools, it gives out 225kJ of heat energy. What is the final temperature of the aluminium?

m = 5.0kg
c = 900J/kg/°C
E = 225kJ = 225 x 10^3J = 225,000J
Θ = ?

The first thing that we will do is calculate the temperature change. Once we have done this we can calculate the final temperature.

$$E = m \times c \times \theta$$

$$\theta = \frac{E}{m \times c}$$

$$\theta = \frac{225{,}000}{5.0 \times 900}$$

(remember to calculate the '5.0 x 900' first as mentioned in example 2)

$$\theta = \frac{225{,}000}{4500}$$

$$\theta = 50°C$$

We have calculated the **temperature change**. If the initial temperature was 70°C and the aluminium has cooled by 50°C, then the final temperature will be:

$$70 - 50 = 20°C.$$

Therefore, the final answer is 20°C.

Example 4

Paraffin wax of mass 2.0kg is supplied with 58×10^3J of heat energy. If the paraffin wax rises in temperature from 10°C to 20°C, calculate the specific heat capacity of paraffin wax.

m = 2.0kg
E = 58×10^3J = 58,000J
θ = 20 - 10 = 10°C
c = ?

$$E = m \times c \times \theta$$

$$c = \frac{E}{m \times \theta}$$

$$c = \frac{58{,}000}{2.0 \times 10}$$

$$c = \frac{58{,}000}{20}$$

c = 2900J/kg/°C

Questions

1) Copper has a specific heat capacity of 385J/kg/°C. If the mass of copper is 4.000kg, calculate the energy supplied to the copper if the temperature change is:
   a) 10.0°C b) 15.0°C c) 50.0°C d) 75.0°C e) 100.0°C
2) Water has a specific heat capacity of 4200J/kg/°C. If 10kg of water are given 2.1MJ of heat energy, calculate the change in temperature.
3) A piece of gold of mass 100g is given 198J of heat energy. If the gold rises in temperature by 15°C, calculate the specific heat capacity of gold.

4) A diamond of specific heat capacity 525J/kg/°C is given 3150J of heat energy. If the diamond rises in temperature by 30°C, calculate the mass of the diamond.
5) A stainless steel block of mass 20.0kg and specific heat capacity 510J/kg/°C rises in temperature by 30.0°C. Calculate the energy supplied.
6) Water has a specific heat capacity of 4200J/kg/°C. If 1000kg of water are given 0.420GJ of heat energy, calculate the change in temperature.
7) Concrete of mass 2000kg is given 33.5×10^6 J of heat energy. If the concrete increases in temperature by 5.0°C, calculate the specific heat capacity of concrete.

29. Latent heat

Latent heat of fusion

When a liquid, for example water at 20°C, is heated to 50°C, the kinetic (movement) energy of the water molecules increases. The greater the kinetic energy of the molecules, the greater the temperature. However, when heat is supplied to a substance the temperature doesn't always increase.
This is the case during a change of state, for example from a solid to a liquid or a liquid to a gas. In a solid, the intermolecular forces (forces between molecules) that hold the molecules together need to be overcome so that they can move freely like in a liquid. When a solid is heated, the vibratory motion of the molecules increases, causing the molecules to move further apart and therefore their potential energy increases. On average, there is no increase in the kinetic energy of the molecules, just an increase in potential energy. As there is no increase in the kinetic energy of the particles, there is no increase in temperature when changing state from a solid to a liquid. Going back to 'work done = energy transferred' in section 14, the energy transferred to the substance while changing state from a solid to a liquid is equal to the work done in separating the molecules, which is why the potential energy is increased.

The specific latent heat of fusion (L_F) of a substance is the amount of heat energy required to change 1kg of a substance from a solid to a liquid without a change in temperature.

The formula for this is given below:

$$E = m \times L_F$$

The amount of heat energy required, E, is measured in joules, J.
The mass, m, is measured in kilograms, kg.
The specific latent heat of fusion, L_F, is measured in joules per kilogram, J/kg.

It is worth noting that this formula can also be used when a liquid changes to a solid without a change in temperature, but energy would have to be removed, not supplied.

Taking a look at rearranging the formula, $E = m \times L_F$, if we want to make **L_F**, the subject of the formula, we need to divide both sides by **m**. This gives:

$$\frac{E}{m} = L_F$$

Therefore,

$$L_F = \frac{E}{m}$$

If we want to make **m** the subject of the formula, going back to the original formula, $E = m \times L_F$, dividing both sides by L_F gives:

$$\frac{E}{L_F} = m$$

Therefore,

$$m = \frac{E}{L_F}$$

Latent heat of vaporization

The specific latent heat of vaporization (L_V) of a substance is the amount of heat energy required to change 1kg of a substance from a liquid to a gas without a change in temperature.

When a liquid changes to a gas, the liquid molecules again have to overcome intermolecular forces if they are to break free so that they can move around freely like in a gas. When a liquid is heated, the vibratory motion of the molecules increases, causing the molecules to move even further apart and their potential energy is increased. On average, there is no increase in the kinetic energy of the molecules, just an increase in potential energy. As there is no increase in kinetic energy of the particles, there is no increase in temperature when changing state from a liquid to a gas. Going back to 'work done = energy transferred' in section 14, the energy transferred to the substance while changing state from a liquid to a gas is equal to the work done in separating the molecules, which is why the potential energy is increased. In the case of changing a liquid to a gas, the molecules are separated so much that the forces between molecules are negligible. The gas molecules are then free to move in the space available to them at high speed.

The formula for the specific latent heat of vaporization is given below:

$$E = m \times L_V$$

The amount of heat energy required, E, is measured in joules, J.
The mass, m, is measured in kilograms, kg.
The specific latent heat of vaporization, L_V, is measured in joules per kilogram, J/kg.

It is worth noting that this formula can also be used when a gas changes to a liquid without a change in temperature, but energy would have to be removed, not supplied.

Taking a look at rearranging the formula, $E = m \times L_V$, if we want to make **L_V**, the subject of the formula, we need to divide both sides by **m**. This gives:

$$\frac{E}{m} = L_V$$

Therefore,

$$L_v = \frac{E}{m}$$

If we want to make **m** the subject of the formula, going back to the original formula, E = m x L_v, dividing both sides by **L_v** gives:

$$\frac{E}{L_v} = m$$

Therefore,

$$m = \frac{E}{L_v}$$

Example 1

Calculate the energy required to change 5.00kg of water at 100°C to steam at 100°C if the specific latent heat of vaporization of water is 226 x 10^4 J/kg.

m = 5.00kg
L_v = 226 x 10^4 J/kg = 2,260,000J/kg
E = ?

$$E = m \times L_v$$

$$E = 5.0 \times 2,260,000$$

$$E = 1.13 \times 10^7 J$$

Or,

$$E = 11,300,000 J$$

Example 2

Calculate the energy required to change 500g of ice at 0°C to water at 0°C if the specific latent heat of fusion for ice is 334kJ/kg.

m = 500g = 0.500kg (remember to convert g to kg divide by 1000)
L_F = 334kJ/kg = 334 x 10^3 J/kg = 334,000J/kg
E = ?

$$E = m \times L_v$$

$$E = 0.500 \times 334,000$$

$$E = 167,000 J$$

Or,

$$E = 1.67 \times 10^5 J$$

Example 3

The specific latent heat of fusion for water is 334kJ/kg. Calculate the mass of water at 0°C changed into ice at 0°C, if the energy that needs to be removed from the water is 668kJ.

m = ?
L_F = 334kJ/kg = 334 x 10^3 J/kg = 334,000J/kg

E = 668kJ = 668 x 10³J = 668,000J

$$E = m \times L_v$$

$$m = \frac{E}{L_F}$$

$$m = \frac{668,000}{334,000}$$

$$m = 2kg$$

Example 4

Calculate the energy required to convert 200g of ice at 0°C to steam at 100°C.
specific latent heat of fusion of ice = 334kJ/kg
specific heat capacity of water = 4200J/kg/°C
specific latent heat of vaporization of water = 2260kJ/kg

This problem needs to be done in 4 stages:
1) Calculate the energy required to change 200g of ice at 0°C to 200g of water at 0°C.
2) Calculate the energy required to change 200g of water at 0°C to water at 100°C (for this we will need the specific heat capacity formula from section 28).
3) Calculate the energy required to change 200g of water at 100°C to 200g of steam at 100°C.
4) Add together the 3 energy values, to get the total amount of energy required.

Step 1

m = 200g = 0.200kg (remember to convert from **g** to **kg**, divide by 1000)
L_F = 334kJ/kg = 334 x 10³J/kg = 334,000J/kg
E = ?

$$E = m \times L_F$$

$$E = 0.200 \times 334,000$$

$$E = 66,800J$$

Step 2

m = 200g = 0.200kg (to convert **g** to **kg** divide by 1000)
c = 4200J/kg/°C
θ = temperature change = 100 − 0 = 100°C
E = ?

$$E = m \times c \times \theta$$

$$E = 0.200 \times 4200 \times 100$$

$$E = 84,000J$$

Step 3

m = 200g = 0.200kg
L$_v$ = 2260kJ/kg = 2260 x 10^{3}J/kg = 2,260,000J/kg
E = ?

$$E = m \times L_v$$

$$E = 0.200 \times 2,260,000$$

$$E = 452,000J$$

Step 4

Adding the 3 energies together we get:

$$\text{Total energy} = 66,800 + 84,000 + 452,000$$

$$\text{Total energy} = 602,800J$$

$$\text{Total energy} = 603kJ \text{ (3 significant figures)}$$

Or,

$$\text{Total energy} = 603,000J$$

Questions

For the questions below, use the following values:

specific latent heat of fusion of ice = 334kJ/kg
specific heat capacity of water = 4200J/kg/°C
specific latent heat of vaporization of water = 2260kJ/kg

1) Calculate the energy required to change 4.000kg of ice at 0°C to water at 0°C.
2) Calculate the energy required to change 2.000kg of water at 100°C to steam at 100°C.
3) Calculate the mass of water at 0°C changed into ice at 0°C, if the energy removed from the water is 133.6kJ.
4) Calculate the energy that needs to be removed to change 500g of water at 0°C to ice at 0°C.
5) Calculate the specific latent heat of fusion of aluminium if 152,000J of energy is needed to change 0.400kg from solid at 659°C to liquid at 659°C.
6) Calculate the energy that needs to be removed from steam to change 2.000kg of steam at 100°C to water at 100°C.
7) Calculate the energy required to convert 800g of ice at 0°C to steam at 100°C.
8) Calculate the energy that needs to be removed to change 50g of steam at 100°C to ice at 0°C.
9) Calculate the energy required to change 10.00g gold at 1067°C from solid to liquid at 1067°C if the specific latent heat of fusion of gold is 7.00 x 10^{4}J/kg.
10) Calculate the energy required to convert 150g of ice at 0°C to steam at 100°C.

30. V = I x R

The formula for potential difference (or voltage) is given below:

$$potential\ difference = current \times resistance$$

Potential difference is measured in volts, V (please note that potential difference, sometimes abbreviated to p.d., is another word for voltage. Therefore, potential difference, voltage and p.d. all mean the same thing).
Current is measured in amperes (or amps), A.
Resistance is measured in ohms, Ω.

This can be written as:

$$V = I \times R$$

Where **V** is the symbol for potential difference (or voltage), **I** is the symbol for current and **R** is the symbol for resistance.

Taking a look at rearranging the formula, $V = I \times R$, if we want to make **I** the subject of the formula, we need to divide both sides by **R**. This gives:

$$\frac{V}{R} = I$$

Therefore,

$$I = \frac{V}{R}$$

Going back to the original formula, $V = I \times R$, if we want to make **R** the subject of the formula, we need to divide both sides by **I**. This gives:

$$\frac{V}{I} = R$$

Therefore,

$$R = \frac{V}{I}$$

In the examples, you will need to know the following circuit symbols:

The symbol for a cell is:

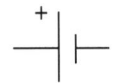

The longer of the 2 vertical lines is the positive terminal, as marked, and the shorter one is the negative terminal of the cell.

In everyday life a cell is known as a battery, but strictly speaking a battery is several cells connected together.

The symbol for a resistor is:

Example 1

Figure 1

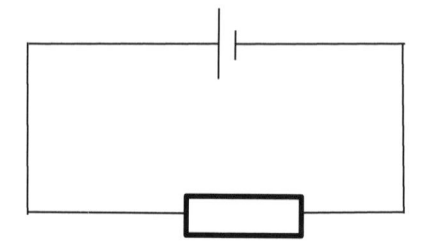

In figure 1 above, calculate the voltage of the cell if the resistor has a resistance of 10Ω and the current in the circuit is 2A.

V = ?
R = 10Ω
I = 2A

$$V = I \times R$$

$$V = 2 \times 10$$

$$V = 20V$$

Example 2

In figure 1 above, calculate the resistance of the resistor, if the cell has a voltage of 6V and the current flowing is 2.0mA.

V = 6V
I = 2.0mA = 2.0 x 10⁻³A = 0.0020A
R = ?

$$V = I \times R$$

$$R = \frac{V}{I}$$

$$R = \frac{6}{0.0020}$$

$$R = 3000 \Omega$$

Example 3

In figure 1 above, calculate the current flowing in the circuit, if the cell has a voltage of 15V and the resistor has a resistance of 5.0kΩ.

V = 15V
R = 5.0kΩ = 5.0 x 10³Ω = 5000Ω
I = ?

$$V = I \times R$$

$$I = \frac{V}{R}$$

$$I = \frac{15}{5000}$$

$$I = 3.0 \times 10^{-3} A$$

Or,

$$I = 3.0 mA \ (0.0030A)$$

Questions

All of the questions below refer to figure 2.

Figure 2

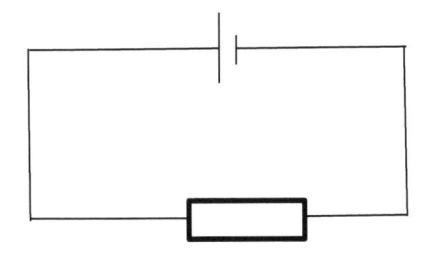

1) Calculate the voltage of the cell if the current flowing in the circuit is 2A and the resistor has a value of 20Ω.
2) Calculate the current in the circuit if the cell has a voltage of 20V and the resistor has a value of 4Ω.
3) Calculate the resistance of the resistor if the current in the circuit is 2.0mA and the cell has a voltage of 4V.
4) Calculate the voltage of the cell if the current flowing in the circuit is 3×10^{-3}A and the resistor has a value of 2000Ω.
5) Calculate the current in the circuit if the cell has a voltage of 50.0V and the resistor has a value of 4.00MΩ.
6) Calculate the resistance of the resistor if the current in the circuit is 4.0A and the cell has a voltage of 100V.
7) Calculate the voltage of the cell if the current flowing in the circuit is 1.5A and the resistor has a value of 4Ω.
8) Calculate the current in the circuit if the cell has a voltage of 25V and the resistor has a value of 5.0kΩ.
9) Calculate the resistance of the resistor if the current in the circuit is 10mA and the cell has a voltage of 35V.
10) Calculate the resistance of the resistor if the current in the circuit is 5.0×10^{-6}A and the cell has a voltage of 40V.

31. Series circuits

A series circuit is one where if you start at the positive terminal of the cell, the current will follow only one path through the electronic components of the

circuit until it reaches the negative terminal of the cell. For example, if you look at figure 1 below, if you start at the positive terminal of the cell, the current will flow through both resistors and back to the negative terminal.
The circuit can take any shape, as long as there is only one path for the current to take.

There are 2 main rules for a series circuit:
1) The current in a series circuit is the same at any point in the circuit.
2) The sum of the voltages across the components (for example resistors) is the same as the voltage across the cell (or battery). As will be shown later in the section, this can be written as:

$$V = V_1 + V_2 + ...$$

To illustrate rule 2, see figure 1 below.

Figure 1

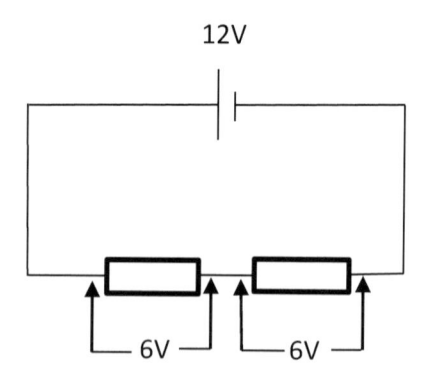

```
                        12V
                        ─┤├─
   ┌─────────────────────────────────────┐
   │                                     │
   │                                     │
   │                                     │
   │         ┌──────┐    ┌──────┐        │
   └─────────┤      ├────┤      ├────────┘
             └──────┘    └──────┘
              ── 6V ──   ── 6V ──
```

In figure 1, the voltage across each resistor is 6V. If we apply rule 2, the sum of the voltages across the components (in this case resistors) is equal to '6 + 6 = 12V'. This is equal to the voltage of the cell which is 12V.

It is important to realise that the voltages across the components don't have to be the same (both are 6V) as they are in this example. They depend on the resistances of the components as will be shown in example 1.
Just to make sure that you understand rule 2, I will give another example in figure 2:

In this example, the voltage across each resistor is 8V and 12V respectively. If we apply rule 2, the sum of the voltages across the components (in this case resistors) is equal to '8 + 12 = 20V'. This is equal to the voltage of the cell which is 20V. It is possible to express rule 2 mathematically. If we let the cell voltage of 20V equal **V**, the 8V equal **V₁** and the 12V equal **V₂**, then we have the formula **V = V₁ + V₂ +**…..

I have put the dots at the end of the formula because if there were 3 resistors in the circuit, you would add the voltages across the 3 resistors together to give V = V₁ + V₂ + V₃. If there were 5 resistors, you would add the voltages across the 5 resistors together to give V = V₁ + V₂ + V₃ + V₄ + V₅.

Figure 2

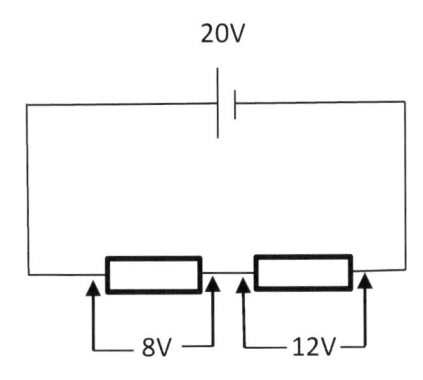

[Circuit diagram: 20V cell with two resistors showing 8V and 12V drops]

Cells connected in series

Figure 3

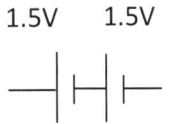

[Diagram of two 1.5V cells in series]

If cells are connected in series as shown in figure 3 above, to find the total voltage supplied by the cells you simply add up the voltage of each cell. Looking at figure 3 above, the total voltage, $V_T = 1.5 + 1.5 = 3.0V$.
It does not matter how many cells are connected in series, you just add together the voltage of each one to find the total voltage (this only applies if the cells are connected correctly as in figure 3 above. See note below).

Note:
If one the cells in figure 3 was reversed in direction, the total voltage, $V_T = 1.5 - 1.5 = 0V$.

Resistors connected in series

Figure 4

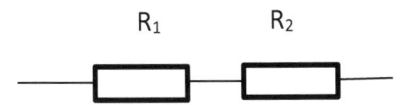

[Diagram of two resistors R_1 and R_2 in series]

When resistors are connected in series the formula to calculate the total resistance is given below:

$$\text{Total resistance, } R_T = R_1 + R_2 + \ldots$$

The formula shows that you just have to add together the resistances to find the total resistance. I have put the dots at the end of the formula because if there were 3 resistors you would add the value of the 3 resistors together. If there were 10 resistors, you would add the value of the 10 resistors together. Therefore in figure 4, if $R_1 = 2Ω$ and $R_2 = 3Ω$ then the total resistance will be:

$$R_T = R_1 + R_2$$

$$R_T = 2 + 3 = 5Ω$$

Example 1

In the circuit below in figure 5 calculate:
a) The total resistance.
b) The current in the circuit.
c) The voltage across each resistor.

Figure 5

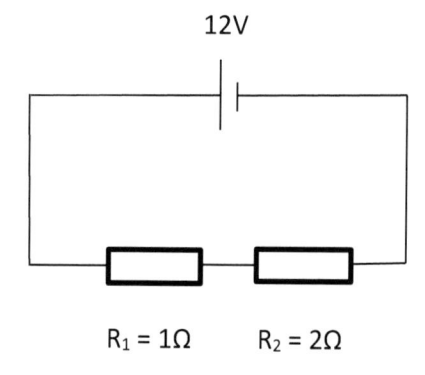

12V

$R_1 = 1Ω$ $R_2 = 2Ω$

a) The total resistance can be calculated using the formula:

$$R_T = R_1 + R_2$$

$$R_T = 1 + 2 = 3Ω$$

b) To calculate the current flowing in the circuit we need to use 'V = I x R' from section 30.

$$V = I \times R$$

We must use the total resistance though as the current has to flow through both resistors.

I = ?
$R_T = 3Ω$
V = 12V

$$I = \frac{V}{R_T}$$

$$I = \frac{12}{3}$$

$$I = 4A$$

c) To calculate the voltage across each resistor we need to consider each resistor individually. We then have to use 'V = I x R'. We also know from rule 1, that the current is the same at any point in the circuit. This means that there must be 4A flowing through each resistor.

Voltage across R_1

Let the voltage across R_1 equal V_1,

$V_1 = ?$
$R_1 = 1\Omega$
$I = 4A$

Therefore,
$$V = I \times R$$

$$V_1 = I \times R_1$$

$$V_1 = 4 \times 1$$

$$V_1 = 4V$$

Voltage across R_2

Let the voltage across R_2 equal V_2,

$V_2 = ?$
$R_2 = 2\Omega$
$I = 4A$

Therefore,
$$V = I \times R$$

$$V_2 = I \times R_2$$

$$V_2 = 4 \times 2$$

$$V_2 = 8V$$

Alternative method for calculating the voltage across R_2

Once V_1 is known, V_2 can be calculated as above or we can use the series circuit rule 2 formula, **V = V$_1$ + V$_2$ +......** . In this example it will be:

$$V = V_1 + V_2$$

but,

$V = 12V$
$V_1 = 4V$
$V_2 = ?$

$$12 = 4 + V_2$$

To make **V$_2$** the subject of the formula we need to subtract 4 from both sides because the '4 and V₂' are added together, so we do the 'opposite' or 'inverse' which is subtract.

$$12 - 4 = V_2$$

Therefore,

$$V_2 = 12 - 4 = 8V$$

Questions

Figure 6

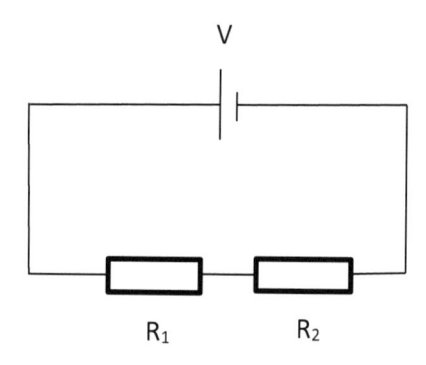

1) In the circuit above in figure 6 calculate:
   a) The total resistance.
   b) The current in the circuit.
   c) The voltage across each resistor, if $V = 20V$, $R_1 = 2\Omega$, $R_2 = 3\Omega$.
2) In the circuit above in figure 6 calculate:
   a) The total resistance.
   b) The current in the circuit.
   c) The voltage across each resistor, if $V = 30V$, $R_1 = 4\Omega$, $R_2 = 6\Omega$.
3) In the circuit above in figure 6 calculate:
   a) The total resistance.
   b) The current in the circuit.
   c) The voltage across each resistor, if $V = 15V$, $R_1 = 2k\Omega$, $R_2 = 3k\Omega$.
4) A series circuit comprises of a cell of voltage 6V and three resistors. $R_1 = 1\Omega$, $R_2 = 2\Omega$ and $R_3 = 3\Omega$. Calculate:
   a) The total resistance.
   b) The current in the circuit.
   c) The voltage across each resistor.
5) A series circuit comprises of a cell of voltage 24V and four resistors. $R_1 = 2\Omega$, $R_2 = 2\Omega$, $R_3 = 3\Omega$ and $R_4 = 5\Omega$. Calculate:
   a) The total resistance.
   b) The current in the circuit.
   c) The voltage across each resistor.

32. Parallel circuits

An example of a parallel circuit is shown below in figure 1:

Figure 1

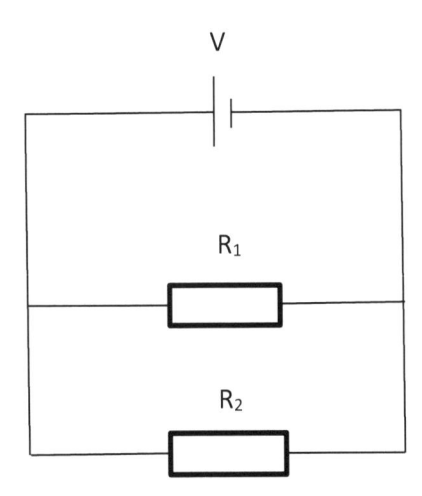

There are 2 rules for parallel circuits:

Rule 1
Any parts of an electrical circuit which are in parallel with each other will have the same voltage across them. In figure 1 above, this means that the voltage of the cell with be the same as the voltage across both R_1 and R_2. For example, if V = 10V, then the voltage across R_1 = 10V and the voltage across R_2 = 10V. This is shown in figure 2 below.

Figure 2

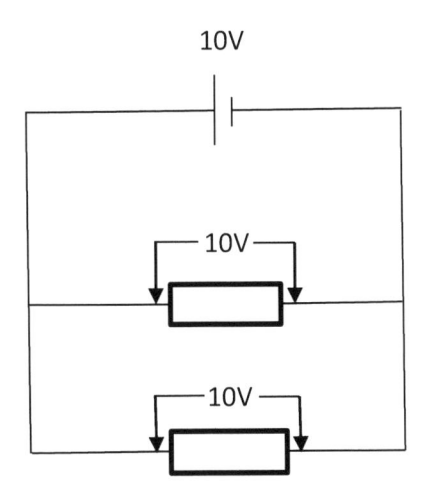

If we let the voltage across R_1 equal V_1 and the voltage across R_2 equal V_2 this can be written mathematically as:

$$V = V_1 = V_2$$

It is important to note that it doesn't matter how many resistors are in parallel, the voltage across each branch would be the same (or in the case of the above example with their only being 1 resistor in each branch, the voltage across each resistor would be the same).

Rule 2

The sum of the currents in the branches of a parallel circuit are equal to the main circuit current (or the current entering or leaving the parallel part of the circuit). This is shown in figure 3 below:

Figure 3

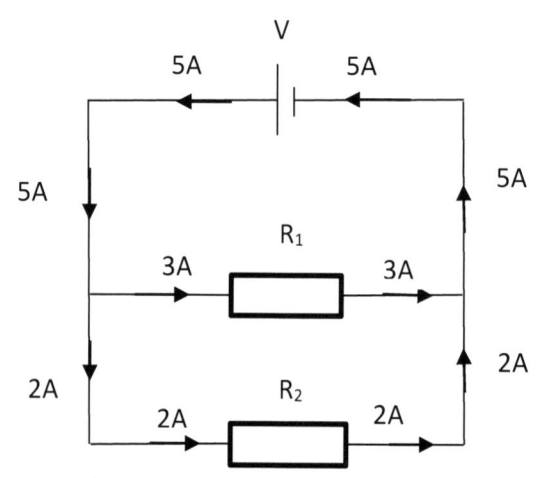

In figure 3 there are 2 branches in the circuit. The branch with R_1 in has a current of 3A flowing in it. The branch with R_2 in has a current of 2A flowing in it. Adding together the currents in the branches of the circuit, we have '2 + 3 = 5A'. This is equal to the main circuit current which is 5A (in figure 3 this is the current through the cell).
As you will see later, the current in each branch depends on the voltage of the cell and the resistance in each branch.

Cells in parallel

If identical cells are connected in parallel as shown in figure 4, their combined voltage is the same as that of 1 of the cells. For example, in figure 4 below, two 1.5V cells are connected in parallel. Their combined voltage is 1.5V.

Figure 4

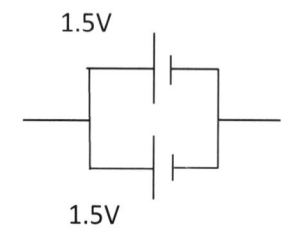

Example 1

In the circuit in figure 5 below calculate:
    a) The voltage across each resistor.
    b) The current through each resistor.
    c) The current through the cell.

Figure 5

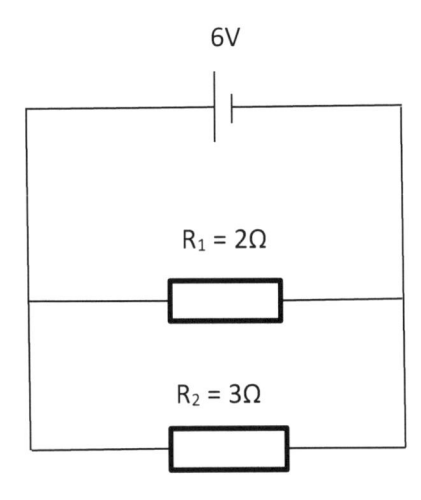

6V

$R_1 = 2\Omega$

$R_2 = 3\Omega$

    a) Applying rule 1, any parts of an electrical circuit which are in parallel with each other will have the same voltage across them. Therefore the voltage across each resistor equals 6V.

    b)

Current through R₁

If we let the voltage across R_1 equal V_1 then:

$V_1 = 6V$
$R_1 = 2\Omega$
$I = ?$

$$V_1 = I \times R_1$$

$$I = \frac{V_1}{R_1}$$

$$I = \frac{6}{2}$$

$$I = 3A$$

Current through R₂

If we let the voltage across R_2 equal V_2 then:

$V_2 = 6V$

R₂ = 3Ω
I = ?

$$V_2 = I \times R_2$$

$$I = \frac{V_2}{R_2}$$

$$I = \frac{6}{3}$$

$$I = 2A$$

c) The current through the cell will according to rule 2, equal the sum of the currents in the branches.

Therefore,
Sum of the current in the branches = 3A + 2A = 5A

Current through cell = 5A

Questions

Figure 6

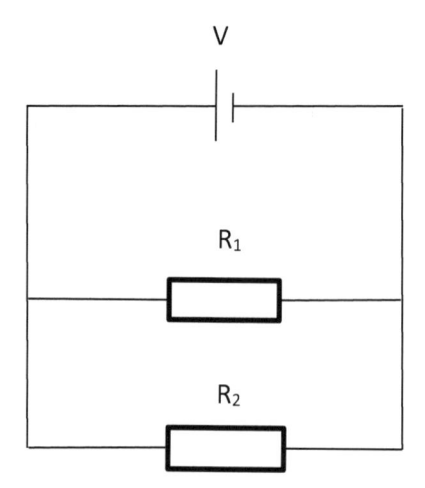

1) In the circuit in figure 6 above calculate:
   a) The voltage across each resistor.
   b) The current through each resistor.
   c) The current through the cell if, V = 2V, R₁ = 4Ω and R₂ = 0.5Ω.
2) In the circuit in figure 6 above calculate:
   a) The voltage across each resistor.
   b) The current through each resistor.
   c) The current through the cell if, V = 10V, R₁ = 2Ω and R₂ = 2Ω.
3) In the circuit in figure 6 above calculate:
   a) The voltage across each resistor.
   b) The current through each resistor.

c) The current through the cell if, V = 8V, R₁ = 1kΩ and R₂ = 2kΩ.
4) A parallel circuit has a cell of 30V with 3 resistors in parallel. Calculate:
a) The voltage across each resistor.
b) The current through each resistor.
c) The current through the cell if, R₁ = 1kΩ, R₂ = 5kΩ and R₃ = 10kΩ.
5) A parallel circuit has a cell of 10V with 3 resistors in parallel. Calculate:
a) The voltage across each resistor.
b) The current through each resistor.
c) The current through the cell if, R₁ = 1kΩ, R₂ = 5kΩ and R₃ = 2kΩ.
6) A student has two 3V cells. Calculate the voltage from the cells if they are connected in a) series (like those connected in section 31, figure 3) b) parallel (like those connected in figure 4 in this section).

33. Q = I x t

The charge that passes a particular point in a circuit in a time, **t**, can be calculated using the formula below:

$$Charge = current \times time$$

Charge is measured in coulombs, C.
Current is measured in amperes (or amps), A.
Time is measured in seconds, s.

This formula can be written as:

$$Q = I \times t$$

Where **Q** is the symbol for charge, **I** is the symbol for current and **t** is the symbol for time.
Let's take a look at rearranging the formula. If we want to make **I** the subject of the formula, we need to divide both sides by **t**. This gives:

$$\frac{Q}{t} = I$$

Therefore,

$$I = \frac{Q}{t}$$

It is from this formula that we get our definition of an electric current. As you can see, current equals charge divided by time. This can be stated as:

Current is the rate of flow of charge

In an electric circuit, the current is due to the flow of electrons. Electrons have a negative charge and flow from the negative terminal of the cell to the positive terminal. The electrons get repelled from the negative terminal and attracted to the positive terminal.
As stated electrons have a negative charge. The magnitude (size) of this charge is 1.6×10^{-19}C. 1 coulomb of charge therefore is comprised of 6.25×10^{18} electrons!

Note:
During the early 1800's when the first cells were manufactured it was thought that current flowed from positive to negative and electrical problems were

solved with current flowing this way around the circuit. It was only later that it was realised that current flowed from negative to positive but it had become the 'convention' to solve problems with the current flowing from positive to negative. Physicists could not be bothered to change the convention so even today when solving electrical problems, current is shown to flow from positive to negative. This is known as 'conventional current'.

So, 'conventional current' flows from positive to negative.
'Electron flow' is from negative to positive.

Going back to the formula, $Q = I \times t$, if we want to make **t** the subject of the formula we need to divide both sides by **I**. This gives:

$$\frac{Q}{I} = t$$

Therefore,

$$t = \frac{Q}{I}$$

Example 1

A current of 2A flows past a particular point in a circuit for 5s. Calculate the charge that passes.

I = 2A
t = 5s
Q = ?

$$Q = I \times t$$
$$Q = 2 \times 5$$
$$Q = 10C$$

Example 2

A charge of 240mC passes a particular point in a circuit in 1 minute. Calculate the current.

$Q = 240mC = 240 \times 10^{-3}C = 0.240C$
t = 1 minute = 60s (remember that time must be in seconds)
I = ?

$$Q = I \times t$$
$$I = \frac{Q}{t}$$
$$I = \frac{0.240}{60}$$
$$I = 0.004A$$

Or,
$$I = 4mA \ (4 \times 10^{-3}A)$$

Example 3

A charge of 72.0kC passes a particular point in a circuit. The current flowing is

5.00A. How long does it take for this charge to pass?

Q = 72.0kC = 72.0 x 10³C = 72,000C
I = 5.00A
t = ?

$$Q = I \times t$$

$$t = \frac{Q}{I}$$

$$t = \frac{72,000}{5.00}$$

t = 14,400s

Note:

1 minute = 60s
1 hour = 60 minutes
Therefore 1 hour = 60 x 60 = 3600s

Dividing 14,400 by 3600 is 4, so 14,400s = 4 hours.

Questions

1) A current of 3A flows past a particular point in a circuit. Calculate the charge that passes if the time that the current flows for is:
    a)   1s b) 5s c) 1 minute d) 1 hour e) 5 hours
2) A charge of 540mC passes a particular point in a circuit. Calculate the current if the time it takes for the charge to pass is:
    a)   10s b) 20s c) 1 minute d) 2 minutes e) 30s
3) A charge of 10.8kC passes a particular point in a circuit. The current flowing is 3.00A. How long does it take for this charge to pass?
4) A charge of 0.360MC passes a particular point in a circuit. The current flowing is 10.0A. How long does it take for this charge to pass?
5) A current of 4A flows past a particular point in a circuit for 1 hour. Calculate the charge that passes.
6) A charge of 720 x 10⁻³C passes a particular point in a circuit in 3 minutes. Calculate the current.
7) A charge of 259.2kC passes a particular point in a circuit. The current flowing is 3.00A. How long does it take for this charge to pass?

34. E = V x Q

The energy transferred in a circuit can be calculated using the formula below:

$$Energy\ transferred = potential\ difference \times charge$$

Energy transferred is measured in joules, J.
Potential difference is measured in volts, V (Remember that potential difference (or p.d.) is just another name for voltage).
Charge is measured in coulombs, C.

This formula can be written as:

$$E = V \times Q$$

Where **E** is the energy, **V** is the voltage or potential difference and **Q** is the charge.

Taking a look at rearranging the formula, if we want to make **V** the subject of the formula, we need to divide both sides by **Q**. This will give:

$$\frac{E}{Q} = V$$

Therefore,

$$V = \frac{E}{Q}$$

From this formula, we can see that voltage equals the energy transferred divided by charge. This gives us a definition of voltage.

Voltage (or potential difference) is the energy transferred per unit charge (per coulomb)
Remember that a unit of charge is 1 coulomb.

Looking again at this formula, voltage equals the energy transferred divided by charge. Energy is measured in joules, J and charge in coulombs, C. This means that:

$$1 \text{ volt} = 1 \text{ joule per coulomb}$$

Or,

$$1V = 1J/C$$

What this means is that if we have a battery of voltage 6V, the battery will give 6 joules of energy to every coulomb of charge that passes (6J/C). A 20V battery would give 20 joules of energy to every coulomb of charge that passes (20J/C). Let's look at the application of this to a circuit in figure 1 below:

Figure 1

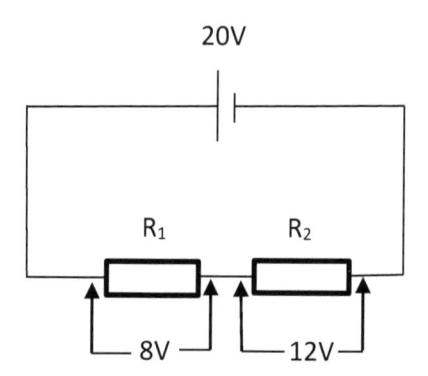

Every coulomb of charge that passes through the 20V cell will be given 20J of energy (20V = 20J/C). Taking the direction of the flow of electrons (negative to positive), if the voltage across R_2 is 12V, then 12J of energy will be given to this resistor and will be given off as heat in the case of a resistor (12V = 12J/C). The voltage across R_1 is 8V and so the remaining 8J will be given to this resistor

(8V = 8J/C). Basically, each coulomb of charge picks up energy from the battery (an amount that depends on the voltage) and this energy is given to electronic components in the circuit that have voltages across them. Whatever the voltage is across a component, each coulomb of charge will give it that amount of energy. So, if a component has a voltage across it of 4V, each coulomb of charge will give it 4J of energy.

Going back to the formula, $E = V \times Q$, if we want to make **Q** the subject of the formula we need to divide both sides by **V**. This gives:

$$\frac{E}{V} = Q$$

Therefore,

$$Q = \frac{E}{V}$$

Example 1

6C of charge passes through a 10V cell. How much energy is transferred to the 6C of charge?

Q = 6C
V = 10V
E = ?

$$E = V \times Q$$

$$E = 10 \times 6$$

E = 60J

Example 2

A battery transfers 42J of energy to 7 coulombs of charge. What is the voltage of the battery?

E = 42J
Q = 7C
V = ?

$$E = V \times Q$$

$$V = \frac{E}{Q}$$

$$V = \frac{42}{7}$$

V = 6V

Example 3

A 12V cell gives out 600J of energy. What is the value of the charge that the 600J is given to?

V = 12V
E = 600J

Q = ?

$$E = V \times Q$$

$$Q = \frac{E}{V}$$

$$Q = \frac{600}{12}$$

$$Q = 50C$$

Questions

1) 8C of charge passes through a cell. How much energy is transferred to the 8C of charge if the voltage of the cell is:
   a) 5V b) 10V c) 20V d) 50V e) 9V
2) A battery transfers 100J of energy. What is the voltage of the battery if this energy is transferred to the following amounts of charge:
   a) 10C b) 5C c) 20C d) 50C e) 25C
3) A cell has a voltage of 9V. What is the value of the charge that is given the following amounts of energy:
   a) 18J b) 90J c) 270J d) 630J e) 540J
4) 10C of charge passes through a 1.5V cell. How much energy is transferred to the 10C of charge?
5) A battery transfers 36J of energy to 4 coulombs of charge. What is the voltage of the battery?
6) A 25V cell gives out 1.0kJ of energy. What is the value of the charge that the 1.0kJ is given to?
7) A battery transfers 3×10^4J of energy to 5kC of charge. What is the voltage of the battery?

35. P = V x I

The formula for electrical power is given below:

$$Electrical\ power = potential\ difference \times current$$

Electrical power is measured in watts, W.
Potential difference is measured in volts, V (Remember that potential difference (or p.d.) is just another name for voltage).
Current is measured in amps, A.

The formula can be written as:

$$P = V \times I$$

Where **P** is the symbol for electrical power, **V** is the symbol for potential difference (or voltage) and **I** is the symbol for current.

As you will see in section 36, another formula for power is:

$$Power = \frac{energy\ transferred}{time}$$

In this formula, energy transferred is measured in joules, J and time is measured in seconds, s. As we are dividing energy by time, if we take a look at the units we are dividing joules by seconds. This means that the unit of power, the watt, W, is equal to the joule per second.

$$1W = 1J/s$$

What this means, for example, is that if we have a 60W light bulb, it will transfer 60J of energy each second. If we have a 100W light bulb, it will transfer 100J of energy each second. Also, looking at the formula, power equals energy transferred divided by time. Another way of wording this is:

power is the rate of transferring energy

Taking a look at rearranging the formula $P = V \times I$, if we want to make **V** the subject of the formula we need to divide both sides by **I**. This gives:

$$\frac{P}{I} = V$$

Therefore,

$$V = \frac{P}{I}$$

Going back to the original formula, $P = V \times I$, if we want to make **I** the subject of the formula we need to divide both sides by **V**. This gives:

$$\frac{P}{V} = I$$

Therefore,

$$I = \frac{P}{V}$$

Example 1

A 2.0kW electric heater is connected to a 230V mains supply. Calculate the current.

P = 2.0kW = 2.0 x 10³W = 2000W (remember that in the formula, power is measured in watts, W).
V = 230V
I = ?

$$P = V \times I$$

$$I = \frac{P}{V}$$

$$I = \frac{2000}{230}$$

I = 8.7A

Example 2

A bulb is rated 12V 3A. Calculate the power of the bulb.

V = 12V
I = 3A

P = ?

$$P = V \times I$$
$$P = 12 \times 3$$
$$P = 36W$$

Example 3

A window fan has a power of 200W. If the current is 0.870A. calculate the voltage.

P = 200W
I = 0.870A
V = ?

$$P = V \times I$$
$$V = \frac{P}{I}$$
$$V = \frac{200}{0.870}$$
$$V = 229.89V$$
$$V = 230V$$

Note:
If you look back to section 30, we looked at the formula, $V = I \times R$. We can combine this with the formula $P = V \times I$, to get some other useful formulae for power.

In the $P = V \times I$ formula, if we replace the **V** with '**I x R**' (because V = I x R), we get:

$$P = I \times R \times I$$

This simplifies to,
P = I²R

Also, with the formula $P = V \times I$, if we replace the **I** with '**V/R**', because I = V/R, we get:

$$P = \frac{V \times V}{R}$$

$$P = \frac{V^2}{R}$$

Example 4

Calculate the power dissipated (given out) by a resistor of size 6.0Ω when a current of 0.40A flows through it.

R = 6.0Ω
I = 0.40A
P = ?

$$P = I^2R$$
$$P = 0.40^2 \times 6.0$$
$$P = 0.16 \times 6.0$$
$$P = 0.96W$$

Questions

1) Calculate the power of a bulb rated 6V 2A.
2) A 6.8kW electric shower is supplied by a voltage of 230V. Calculate the current.
3) A microwave is supplied by a 230V supply. If the current flowing is 4.35A, calculate the power of the microwave.
4) A 0.800kW toaster is supplied by a voltage of 230V. Calculate the current.
5) A LCD TV is rated 230V 0.65A. Calculate the rate at which energy is transferred.
6) A 900W electric drill is supplied by a voltage of 230V. Calculate the current.
7) A 2.20kW kettle is supplied by a voltage of 230V. Calculate the current.
8) Calculate the power dissipated by a resistor of size 10Ω when a current of 0.50A flows through it.
9) Calculate the power dissipated by a resistor of size 25Ω when the voltage across it is 5V.
10) Calculate the current that flows through a bulb rated 60W 230V.

36. E = P x t

The formula for energy transferred is given below:

$$Energy\ transferred = Power \times time$$

Energy is measured in joules, J.
Power is measured in watts, W.
Time is measured in seconds, s.

This can be written as:

$$E = P \times t$$

Taking a look at rearranging the formula, if we want to make **P** the subject of the formula we need to divide both sides by **t**. This will give:

$$\frac{E}{t} = P$$

Therefore,

$$P = \frac{E}{t}$$

In this formula, energy transferred is measured in joules, J and time is measured in seconds, s. As we are dividing energy by time, if we take a look at the units we are dividing joules by seconds. This means that the unit of power, the watt, W, is equal to the joule per second.

$$1W = 1J/s$$

What this means, for example, is that if we have a 60W light bulb, it will transfer 60J of energy each second. If we have a 100W light bulb, it will transfer 100J of energy each second. Also, looking at the formula, power equals energy transferred divided by time. Another way of wording this is:

$$power\ is\ the\ rate\ of\ transferring\ energy$$

Going back to the formula, $E = P \times t$, if we want to make **t** the subject of the formula we need to divide both sides by **P**. This gives:

$$\frac{E}{P} = t$$

Therefore,

$$t = \frac{E}{P}$$

Note:
From section 14, **work done = energy transferred**, so, in all of the above formulae we can replace 'energy transferred' with 'work done' to get the following formulae:

$$W = P \times t$$

$$P = \frac{W}{t}$$

$$t = \frac{W}{P}$$

Where **W** is equal to the 'work done'.

Example 1

An 800W microwave is switched on for 2 minutes. Calculate the electrical energy supplied to the microwave.

P = 800W
t = 2 minutes = 120s (remember that time must be in seconds)
E = ?

$$E = P \times t$$

$$E = 800 \times 120$$

$$E = 96,000J$$

Example 2

An electric fire is supplied with 28.8MJ of electrical energy in 4 hours. Calculate the power of the electric fire.

Firstly we need to convert 4 hours into seconds.

1 minute = 60s
60 minutes = 1 hour.
Therefore 1 hour = 60 x 60 = 3600s.
4 hours = 4 x 3600 = 14,400s

E = 28.8MJ = 28.8 x10⁶J = 28,800,000J
t = 4 hours = 14,400s
P = ?

$$E = P \times t$$

$$P = \frac{E}{t}$$

$$P = \frac{28,800,000}{14,400}$$

P = 2000W

Example 3

A hedge trimmer of power 500W is supplied with 900kJ of energy. How long was it switched on for?

P = 500W
E = 900kJ = 900 x 10³J = 900,000J
t = ?

$$E = P \times t$$

$$t = \frac{E}{P}$$

$$t = \frac{900,000}{500}$$

t = 1800s

Questions

1) A dishwasher transfers 5,760,000J of energy while switched on for 80 minutes. Calculate the power of the dishwasher.
2) Calculate the energy transferred by a 100W light bulb if it is switched on for:
   a) 1 minute b) 30 minutes c) 1 hour d) 2 hours e) 10 hours
3) A 3.0kW oven is supplied with 16.2MJ of energy. How long is the oven switched on for?
4) A stereo is supplied with 432kJ of energy in 2 hours. Calculate the power of the stereo.
5) A 0.300kW blender is switched on for 3 minutes. How much energy is supplied to the blender.
6) An electric blanket of power 200W is supplied with 5.76 x 10⁶J of energy. How long is it switched on for?
7) A 500W TV is switched on for 2 hours. How much energy is transferred?

37. Paying for electricity

As seen in section 36, the energies supplied to electrical appliances in joules can be very large. If electrical meters used by energy suppliers were measured in joules, the meter readings would be massive. To avoid this, electricity meters measure energy in kilowatt-hours, rather than joules.
The formula for calculating the energy transferred to electrical appliances when 'paying for electricity' is the same as in section 36 except that different units are used.

$$Energy\ transferred = Power \times time$$

When paying for electricity however,
Energy transferred is measured in the kilowatt-hour, kWh. It is worth noting that energy transferred can also be measured in 'units'. **1 kWh = 1 unit**.
Power is measured in kilowatts, kW.
Time is measured in hours, h.

Once the energy has been calculated (in kWh or units) the cost of electricity can be calculated using the formula below:

$$cost = number\ of\ kWh's\ (or\ units) \times cost\ per\ kWh\ (or\ unit)$$

When calculating the cost of using an electrical appliance, 2 steps are required in the calculation:
Step 1 – Calculate the energy transferred (in kWh or units).
Step 2 – Calculate the cost using the cost formula given above.

Example 1

A 3000W oven is switched on for 2 hours. Calculate the cost, if electricity is 10p/kWh (10 pence per kilowatt-hour).

The first thing that we need to do is convert the 3000W into kilowatts, kW.
To convert W into kW you divide by 1000.
Therefore, 3000W = 3000/1000 = 3kW
P = 3kW
t = 2 hours
E = ?

$$Energy\ transferred = Power \times time$$

$$E = 3 \times 2$$

$$E = 6kWh$$

The cost of electricity is 10p/kWh, therefore:

$$cost = number\ of\ kWh's \times cost\ per\ kWh$$

$$cost = 6 \times 10$$

$$cost = 60p$$

Example 2

A 100W light bulb is switched on for 3 hours. Calculate the cost, if electricity is 8p/unit (8 pence per unit). (remember 1kWh = 1 unit)

The first thing that we need to do is convert the 100W into kilowatts, kW.
To convert W into kW you divide by 1000.
Therefore, 100W = 100/1000 = 0.1kW
P = 0.1kW
t = 3 hours
E = ?

$$Energy\ transferred = Power\ x\ time$$

$$E = 0.1\ x\ 3$$

$$E = 0.3\ units$$

The cost of electricity is 8p/unit, therefore:

$$cost = number\ of\ units\ x\ cost\ per\ unit$$

$$cost = 0.3\ x\ 8$$

$$cost = 2.4p$$

With the questions below, remember that time must be in hours and that there are 60 minutes in 1 hour. For example, let's say that a time of 40 minutes is given in a question. As a fraction of an hour this will be:

$$\frac{40}{60} = \frac{2}{3}\ of\ an\ hour$$

Therefore to convert minutes into hours you must divide by 60!

Questions

1) Calculate the energy transferred to a 40W light bulb if it is switched on for:
   a) 1 hour b) 5 hours c) 10 hours
2) A 100W light bulb is switched on for 4 hours. Calculate the cost of the electricity if it is charged at 10p/unit.
3) A 1500W microwave is switched on for 20 minutes. Calculate the cost of the electricity if it is charged at 10p/unit.
4) A 0.800kW toaster is switched on for 3 minutes. Calculate the cost of the electricity if it is charged at 8p/unit.
5) A 200W TV is switched on for 2 hours. Calculate the cost of the electricity if it is charged at 10p/unit.
6) A 1200W heater is left switched on for 1 day. Calculate the cost of the electricity if it is charged at 10p/kWh.
7) A 1kW electric drill is switched on for 10 minutes. Calculate the cost of the electricity if it is charged at 10p/unit.
8) Five 100W bulbs are switched on for 2 hours. Calculate the cost of the electricity if it is charged at 10p/kWh.
9) A 7000W electric shower is switched on for 15 minutes. Calculate the cost of the electricity if it is charged at 10p/unit.

10) A 2000W kettle is switched on for 180s. Calculate the cost of the electricity if it is charged at 10p/unit.

38. E = V x I x t

Going back to the electrical power section (section 35), **P = V x I**. If we substitute this into the formula **E = P x t** from section 36, replacing the **P** with '**V x I**', we get:

$$E = V \times I \times t$$

E is the energy transferred measured in joules, J.
V is the voltage measured in volts, V.
I is the current measured in amps, A.
t is the time measured in seconds, s.

This formula can then be used to calculate the electrical energy supplied to electrical appliances or components.

Taking a look at rearranging the formula, if we want to make **V** the subject of the formula, we need to divide both sides by '**I x t**' or '**It**'. Remember that the **V**, **I** and **t** are all multiplied together, so to get the **V** by itself you must do the 'opposite' or 'inverse', which is divide. This will give:

$$\frac{E}{I \times t} = V$$

Therefore,

$$V = \frac{E}{I \times t}$$

If we want to make **I** the subject of the formula, going back to the original formula, $E = V \times I \times t$, we need to divide both sides by '**V x t**'. This gives:

$$\frac{E}{V \times t} = I$$

Therefore,

$$I = \frac{E}{V \times t}$$

Finally, if we want to make **t** the subject of the formula, going back to the original formula, $E = V \times I \times t$, we need to divide both sides by '**V x I**'. This will give:

$$\frac{E}{V \times I} = t$$

Therefore,

$$t = \frac{E}{V \times I}$$

Example 1

A resistor has a voltage across it of 2.0V and a current flowing through it of 0.10A for 10 minutes. Calculate the energy supplied to the resistor.

V = 2.0V
I = 0.10A
t = 10 minutes = 600s (Remember that there are 60s in 1 minute. Therefore to convert minutes to seconds, you multiply by 60. So, '10 x 60 = 600', therefore there are 600s in 10 minutes.
E = ?

$$E = V \, x \, I \, x \, t$$

$$E = 2.0 \; x \; 0.10 \; x \; 600$$

$$E = 120J$$

Example 2

An electric fan is switched on for 30 minutes. The energy supplied to it in this time is 372.6kJ and the voltage is 230V. Calculate the current.

t = 30 minutes = 1800s
E = 372.6kJ = 372.6 x 10^3J = 372,600J
V = 230V
I = ?

$$E = V \, x \, I \, x \, t$$

$$I = \frac{E}{V \, x \, t}$$

$$I = \frac{372,600}{230 \; x \; 1800}$$

$$I = \frac{372,600}{414,000}$$

$$I = 0.90A$$

Note:
Please refer back to section 28, example 2 which explains how to avoid an unnecessary mathematical mistake when using a formula of this kind.

Example 3

An electric drill is supplied with 269.1kJ of energy. If the voltage supplied is 230V and the current flowing through it is 3.90A, calculate the time that it is switched on for.

E = 269.1kJ = 269.1 x 10^3J = 269,100J
V = 230V
I = 3.90A
t = ?

$$E = V \, x \, I \, x \, t$$

$$t = \frac{E}{V \times I}$$

$$t = \frac{269,100}{230 \times 3.90}$$

$$t = \frac{269,100}{897}$$

$$t = 300s$$

Questions

1) A resistor has a voltage across it of 4.0V and a current flowing through it of 0.20A for 5 minutes. Calculate the energy supplied to the resistor.
2) A LCD TV is supplied with 1.3455MJ of energy. If the supply voltage is 230V and the current passing through it is 0.65A, how long was the TV switched on for?
3) An electric shower is switched on for 10.0 minutes. If the shower has a supply voltage of 230V and during this time is supplied with 4084800J of energy, calculate the current.
4) A bulb is supplied with 10.8kJ of energy. If it is switched on for 15 minutes and the current flowing through it is 2A, calculate the voltage across the bulb.
5) A toaster has a supply voltage of 230V and a current passing through it of 3.48A. If it is switched on for 120s, calculate the energy supplied to the toaster.
6) A resistor has a voltage across it of 1.5V and a current flowing through it of 10mA. If the energy supplied to the resistor is 2.7J, how long did it take to supply this amount of energy.
7) A dishwasher is supplied with 5.0232×10^6J of energy. If it has a supply voltage of 230V and is switched on for 70 minutes, calculate the current.

39. Efficiency

When a device is supplied with energy, some of the energy is transferred usefully while some of it is wasted. For example, if a light bulb is supplied with 100J of energy, 20J might be converted to light (useful) while 80J would be converted to heat (wasted). This information can be used to calculate the efficiency.

The formula for efficiency is given below:

$$efficiency = \frac{useful\ energy\ output}{total\ energy\ input}$$

Useful energy output is measured in joules, J.
Total energy input is measured in joules, J.
Efficiency has no units and is expressed as a fraction or decimal. It can also be expressed as a percentage, if the fraction or decimal is multiplied by 100, as shown in example 1.

With some devices it is useful to look at the energy each second, rather than just energy. The 'energy per second' is another name for power, so another formula that can be used for efficiency is:

$$efficiency = \frac{useful\ power\ output}{total\ power\ input}$$

I shall rearrange the 'energy' version of the formula, but if you want to use the 'power' version, all you have to do is replace the word 'energy' with 'power'. If you use the power version of the formula, you need to remember that power is measured in watts, W.

Taking a look at the 'energy' version of the efficiency formula, to make **useful energy output** the subject of the formula, we need to multiply both sides by **total energy input**. This gives:

$$efficiency \times total\ energy\ input = useful\ energy\ output$$

Therefore,

$$useful\ energy\ output = efficiency \times total\ energy\ input$$

If we want to make **total energy input** the subject of the formula, from the previous line above we have:

$$useful\ energy\ output = efficiency \times total\ energy\ input$$

If we divide both sides by **efficiency**, this will give:

$$\frac{useful\ energy\ output}{efficiency} = total\ energy\ input$$

Therefore,

$$total\ energy\ input = \frac{useful\ energy\ output}{efficiency}$$

Example 1

A machine is supplied with 100J of energy. If the useful energy output is 60J, calculate the efficiency.

Total energy input = energy supplied = 100J
useful energy output = 60J
efficiency = ?

$$efficiency = \frac{useful\ energy\ output}{total\ energy\ input}$$

$$efficiency = \frac{60}{100}$$

efficiency = 0.6

If you want to convert the efficiency into a percentage, you need to multiply by 100.

$$efficiency = 0.6 \times 100$$

Efficiency = 60%

The efficiency of the machine is therefore 0.6 or 60%.

Example 2

A tungsten filament light bulb has an efficiency of 20%. If the total power input is 40W, calculate the useful power output.

When calculating either the 'useful power output' or the 'useful energy output' the efficiency must be expressed as a fraction or decimal. If in the question, the efficiency has been given as a percentage, it must be converted to a fraction or decimal. To do this, you divide the efficiency expressed as a percentage by 100.

efficiency = 20% = 20/100 = 0.20
total power input = 40W
useful power output = ?

$$efficiency = \frac{useful\ power\ output}{total\ power\ input}$$

$$useful\ power\ output = efficiency \times total\ power\ input$$

$$useful\ power\ output = 0.20 \times 40$$

useful power output = 8W

Example 3

A kettle is 90% efficient. If the useful power output is 1980W, calculate the total power input.

When calculating either the 'total power input' or the 'total energy input' the efficiency must be expressed as a fraction or decimal. If in the question, the efficiency has been given as a percentage, it must be converted to a fraction or decimal. To do this, you divide the efficiency expressed as a percentage by 100.

Efficiency = 90% = 90/100 = 0.90
Useful power output = 1980W
total power input = ?

$$efficiency = \frac{useful\ power\ output}{total\ power\ input}$$

$$total\ power\ input = \frac{useful\ power\ output}{efficiency}$$

$$total\ power\ input = \frac{1980}{0.90}$$

total power input = 2200W

Questions

1) A machine is 75.0% efficient. Calculate the total energy supplied to the machine, if the useful energy output is:
   a) 100J b) 500J c) 1800J d) 3×10^3J e) 4MJ

2) A machine is 40% efficient. Calculate the useful energy output, if the total energy input is:
   a) 100J b) 500J c) 750J d) 6×10^4 J e) 7.0kJ
3) A low energy light bulb is supplied with 2000J/s. If the useful power output is 1500J/s, calculate the efficiency of the bulb. (remember that power is measured in watts, W and that 1W = 1J/s).
4) A machine is has a total power input of 5.00×10^3 W. If it has an efficiency of 35.0%, calculate the useful power output.
5) An engine is supplied with 10.0MJ of energy. If the engine is 35% efficient, calculate the useful energy output.
6) A food mixer is 60% efficient. Calculate the total power input if the useful power output is 600W.
7) A machine has a total energy input of 5.0×10^8 J. If the useful energy output is 1.5×10^8 J, calculate the efficiency of the machine.

40. Transformers

A transformer is a device which consists of a primary and secondary coil that are both wound onto a soft iron core. The transformer 'transforms' or 'changes' an alternating voltage in the primary coil into either a greater or smaller alternating voltage in the secondary coil (this depends on the number of turns on the primary and secondary coils). As we shall see later in this section, it is important to know that it is not just the alternating voltage that changes but also the alternating current. We shall see in detail how they affect each other later in the section.

The formula for the transformer is given below:

$$\frac{V_P}{V_S} = \frac{N_P}{N_S}$$

V_P is the voltage across the primary coil measured in volts, V.
V_S is the voltage across the secondary coil measured in volts, V.
N_P is the number of turns on the primary coil.
N_S is the number of turns on the secondary coil.

Taking a look at the formula, if we want to make V_P the subject of the formula, we need to multiply both sides by V_S. This gives:

$$V_P = \frac{N_P \times V_S}{N_S}$$

Going back to the original formula:

$$\frac{V_P}{V_S} = \frac{N_P}{N_S}$$

If we want to make N_P the subject of the formula, we need to multiply both sides by N_S. This will give:

$$\frac{V_P \times N_S}{V_S} = N_P$$

Therefore,

$$N_P = \frac{V_P \times N_S}{V_S}$$

From the formula that we have just proved which has N_P as the subject of the formula, if we multiply both sides by **V_S** we get:

$$V_S \times N_P = V_P \times N_S$$

If we now want to make **V_S** the subject of the formula, we need to divide both sides by **N_P**. This gives:

$$V_S = \frac{V_P \times N_S}{N_P}$$

Finally, if we want to make **N_S** the subject of the formula, going back to the formula below:

$$V_S \times N_P = V_P \times N_S$$

Dividing both sides by **V_P** gives:

$$\frac{V_S \times N_P}{V_P} = N_S$$

Therefore,

$$N_S = \frac{V_S \times N_P}{V_P}$$

Regarding transformers, there is another important formula that we need to consider. If we assume that the transformer is 100% efficient, this means that the power in the primary coil is equal to the power in the secondary coil. This can be written as:

$$\text{power in primary coil} = \text{power in secondary coil}$$

Looking back to section 35, power, $P = V \times I$. Therefore:

$$V_P \times I_P = V_S \times I_S$$

V_P is the voltage across the primary coil measured in volts, V.
I_P is the current flowing in the primary coil measured in amps, A.
V_S is the voltage across the secondary coil measured in volts, V.
I_S is the current flowing in the secondary coil measured in amps, A.

If we want to make **V_P** the subject of this formula, we need to divide both sides by **I_P**. This gives:

$$V_P = \frac{V_S \times I_S}{I_P}$$

The formula, **$V_P \times I_P = V_S \times I_S$**, can be rearranged in a similar manner to give the following:

$$I_P = \frac{V_S \times I_S}{V_P}$$

$$V_S = \frac{V_P \times I_P}{I_S}$$

Physics calculations for GCSE & IGCSE

$$I_S = \frac{V_P \times I_P}{V_S}$$

I suggest that you attempt making **I_P, V_S** and **I_S** the subject of the formula yourselves to make sure that you can do it.

Note:

A step-up transformer is one where the secondary voltage is bigger than the primary voltage.
A step-down transformer is one where the secondary voltage is smaller than the primary voltage.

Example 1

A transformer has 100 turns on the primary coil and 400 turns on the secondary coil. If the primary coil has a voltage across it of 25V, calculate the voltage across the secondary coil.

N_P = 100 turns
N_S = 400 turns
V_P = 25V
V_S = ?

$$\frac{V_P}{V_S} = \frac{N_P}{N_S}$$

$$V_S = \frac{V_P \times N_S}{N_P}$$

$$V_S = \frac{25 \times 400}{100}$$

$$V_S = \frac{10{,}000}{100}$$

$$V_S = 100V$$

Example 2

Electricity is generated at a power station at 25kV. If the transformer steps the voltage up to 400kV and the current is 250A, calculate the current in the primary coil.

V_P = 25kV = 25 x 10^3 V = 25,000V
V_S = 400kV = 400 x 10^3 V = 400,000V
I_S = 250A
I_P = ?

$$V_P \times I_P = V_S \times I_S$$

$$I_P = \frac{V_S \times I_S}{V_P}$$

$$I_P = \frac{400{,}000 \times 250}{25{,}000}$$

$$I_P = \frac{100{,}000{,}000}{25{,}000}$$

$$I_P = 4{,}000A$$

It is worth noting here that we have a step-up transformer. The voltage has been 'stepped up'. This means that the secondary voltage is bigger than the primary voltage. Note however, that in order to keep the power in the primary and secondary the same, that the secondary current is smaller than the primary current. In a step-up transformer, the voltage is 'stepped up' but the current is 'stepped down'. This is extremely important in the transmission of electricity as the larger the current in a wire, the greater the heating effect in the wire. This increases the energy losses and reduces the efficiency of transmission. The most efficient transmission takes place when currents are small. If you look back to section 35, power loss (energy loss per second) can be calculated by the formula:

$$P = I^2 \times R$$

From this formula, you can see that for a fixed resistance value, the larger the current the larger the power loss.

In more detail, $P \propto I^2$ (power loss is proportional to the current squared). This means, for example, that if the current was doubled, the power loss would be quadrupled (4 times bigger). If the current was tripled the power loss would be 9 times bigger.

To explain this point further let's say that $I = 1A$ and $R = 10\Omega$. From the formula, **$P = I^2 \times R$**, this gives $P = 1^2 \times 10 = 1 \times 10 = 10W$. If we now double the current, $I = 2A$ and $R = 10\Omega$. This will now give, $P = 2^2 \times 10 = 4 \times 10 = 40W$. The current has doubled, but the power has increased so that it is now 4 times bigger!

This example showed that in a step-up transformer, the voltage is 'stepped up' but the current is 'stepped down'. A step-down transformer does the opposite, the voltage is 'stepped down' but the current is 'stepped up'.

Example 3

A transformer has a primary voltage of 230V. If the secondary voltage is 5V and the number of turns on the secondary is 100, calculate the number of turns on the primary.

$V_P = 230V$
$V_S = 5V$
$N_S = 100$ turns
$N_P = ?$

$$\frac{V_P}{V_S} = \frac{N_P}{N_S}$$

$$N_P = \frac{V_P \times N_S}{V_S}$$

$$N_P = \frac{230 \times 100}{5}$$

$$N_P = \frac{23,000}{5}$$

$$N_P = 4600 \text{ turns}$$

Questions

1) A transformer has 50 turns on the primary coil and 250 turns on the secondary coil. If the primary coil has a voltage across it of 20V, calculate the voltage across the secondary coil.
2) A transformer has a primary voltage of 230V. If the secondary voltage is 10V and the number of turns on the secondary is 25, calculate the number of turns on the primary.
3) Electricity is generated at a power station at 25,000V. If the transformer steps the voltage up to 275,000V and the current is 365.0A, calculate the current in the primary coil.
4) A transformer has 120 turns on the secondary coil. If the primary coil has a voltage across it of 5V and the voltage across the secondary coil is 20V, calculate the number of turns on the primary coil.
5) A transformer has 40 turns on the primary coil with a voltage across it of 15V. If the voltage across the secondary coil is 120V, calculate the number of turns on the secondary coil.
6) A transformer has a current through the primary coil of 2.0A. If the current flowing through the secondary coil is 0.50A and the voltage across it is 10V, calculate the voltage across the primary coil.
7) A transformer has a current flowing through the primary coil of 5A and a voltage across it of 20V. If the voltage across the secondary is 50V, calculate the current flowing through the secondary coil.

41. v = f x λ

Before looking at this formula there are a few definitions that I need to explain.

Frequency is defined as the number of complete waves per second. It is measured in hertz, Hz. It can also be defined as the number of complete vibrations/oscillations per second.

Wavelength is the distance occupied by one complete wave. It is measured in metres, m (the symbol used for wavelength is the Greek letter 'lambda', λ). This is shown in figure 1 below:

Figure 1

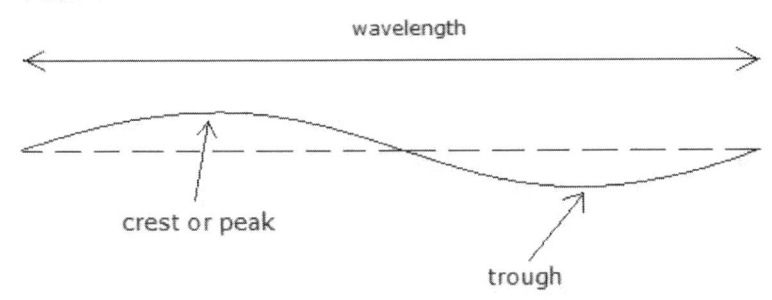

The highest point on the wave is called the crest. The lowest point is called the trough. A useful alternative for the definition of wavelength is that it is the distance between 2 successive crests (the distance between one crest and the next crest) or the distance between 2 successive troughs. This is shown in figure 2.

The formula for wave speed, **v**, is given below:

$$wave\ speed = frequency \times wavelength$$

Wave speed is measured in metres per second, m/s.
Frequency is measured in hertz, Hz.
Wavelength is measured in metres, m.

This formula can be written as:

$$v = f \times \lambda$$

Where **v** is the symbol for wave speed, **f** is the symbol for frequency and **λ** is the symbol for wavelength (**λ**, is the Greek letter 'lambda').

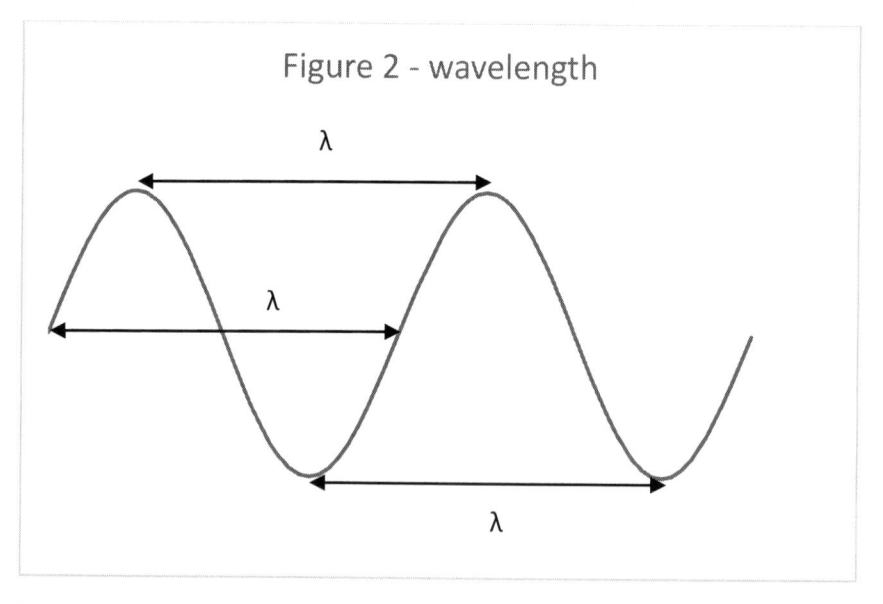

Figure 2 - wavelength

Taking a look at rearranging the formula, $v = f \times \lambda$, if we want to make **f** the subject of the formula we need to divide both sides by **λ**. This gives:

$$\frac{v}{\lambda} = f$$

Therefore,

$$f = \frac{v}{\lambda}$$

If we want to make **λ** the subject of the formula, going back to the original formula, $v = f \times \lambda$, we need to divide both sides by **f**. This will give:

Therefore,
$$\frac{v}{f} = \lambda$$

$$\lambda = \frac{v}{f}$$

Example 1

A wave has a wavelength of 2.0m and a frequency of 10.0Hz. Calculate the wave speed.

λ = 2.0m
f = 10.0Hz
v = ?

$$v = f \times \lambda$$

$$v = 10.0 \times 2.0$$

$$v = 20\text{m/s}$$

Example 2

Free radio Birmingham has a frequency of 96.4MHz. If the speed of radio waves is 3.0×10^8 m/s, calculate the wavelength of the radio waves.

f = 96.4MHz = 96.4×10^6Hz = 96,400,000Hz
v = 3.0×10^8m/s = 300,000,000m/s
λ = ?

$$v = f \times \lambda$$

$$\lambda = \frac{v}{f}$$

$$\lambda = \frac{300,000,000}{96,400,000}$$

$$\lambda = 3.1\text{m}$$

Example 3

Red light has a wavelength of 7.0×10^{-7}m. If the speed of light is 3.0×10^8m/s, calculate the frequency of red light.

λ = 7.0×10^{-7}m = 0.00000070m
v = 3.0×10^8m/s = 300,000,000m/s
f = ?

$$v = f \times \lambda$$

$$f = \frac{v}{\lambda}$$

$$f = \frac{300,000,000}{0.00000070}$$

$$f = 4.3 \times 10^{14} Hz$$

This is equal to 430,000,000,000,000Hz, but it is obviously easier to express this as 4.3×10^{14}Hz.

Questions

1) A wave has a frequency of 10Hz. Calculate the wave speed if the wavelength is:
   a) 1.0m b) 4.0m c) 10m d) 0.5m e) 0.25m
2) A wave has a wave speed of 20m/s. Calculate the frequency if the wavelength is:
   a) 4m b) 500cm c) 10m d) 500mm e) 200cm
   Hint: remember that wavelength must be measured in metres!
3) A wave has a wave speed of 10cm/s. Calculate the wavelength if the frequency is:
   a) 2Hz b) 1Hz c) 5 Hz d) 20Hz e) 100Hz
4) Violet light has a wavelength of 4.0×10^{-7}m. If the speed of light is 300,000,000m/s, calculate the frequency of violet light.
5) Radio 5 live has a frequency of 693kHz. If the speed of radio waves is 3.00×10^{8}m/s, calculate the wavelength of the radio waves.
6) Microwaves have a wavelength of 3.0cm. If the speed of microwaves is 3×10^{8}m/s, calculate the frequency of the microwaves.
7) A wave has a wavelength of 30mm. If the frequency is 5.0Hz, calculate the wave speed.

42. T = 1/f

Time period, T, is the time taken for 1 complete wavelength to pass a particular point. It can also be described as the time taken for 1 complete vibration/oscillation.

The formula for time period, T, is given below:

$$Time\ period = \frac{1}{frequency}$$

Time period is measured in seconds, s.
Frequency is measured in hertz, Hz.

This can be written as:

$$T = \frac{1}{f}$$

Taking a look at rearranging the formula, if we want to make **f** the subject of the formula, the first thing we need to do is multiply both sides by **f**. This gives:

$$T \times f = 1$$

In order to get **f** by itself we need to remove the **T** from the left hand side of the formula. To do this we need to divide both sides by **T**. This is because the '**T and f**' are multiplied together and therefore we do the 'opposite' or 'inverse' to remove it from the left hand side. This gives:

Physics calculations for GCSE & IGCSE

$$f = \frac{1}{T}$$

Example 1

An oscilloscope trace of an a.c. (alternating current) voltage shows that the time period of the trace is 0.02s. Calculate the frequency of the a.c. voltage.

T = 0.02s
f = ?

$$T = \frac{1}{f}$$

$$f = \frac{1}{T}$$

$$f = \frac{1}{0.02}$$

f = 50Hz

Example 2

A pendulum has a frequency of 4Hz. Calculate the time period.

f = 4Hz
T = ?

$$T = \frac{1}{f}$$

$$T = \frac{1}{4}$$

T = 0.25s

Example 3

A wave has a time period of 5ms. Calculate the frequency.

T = 5ms = 5 x 10^{-3}s = 0.005s
f = ?

$$T = \frac{1}{f}$$

$$f = \frac{1}{T}$$

$$f = \frac{1}{0.005}$$

f = 200Hz

Example 4

Free radio Birmingham has a frequency of 96.4MHz. Calculate the time period of the radio wave.

$f = 96.4\text{MHz} = 96.4 \times 10^6 \text{Hz} = 96{,}400{,}000\text{Hz}$
$T = ?$

$$T = \frac{1}{f}$$

$$T = \frac{1}{96{,}400{,}000}$$

$T = 1.04 \times 10^{-8}\text{s}$

Questions

1) A pendulum has a frequency of 10Hz. Calculate the time period.
2) A wave has a time period of 0.1s. Calculate the frequency.
3) A pendulum has a time period of 0.5s. Calculate the frequency.
4) A wave has a frequency of 500Hz. Calculate the time period.
5) A wave has a time period of 20ms. Calculate the frequency.
6) A wave has a frequency of 1kHz. Calculate the time period.
7) Calculate the time period of microwaves with a frequency of 1GHz.
8) Calculate the time period of gamma rays with a frequency of 1×10^{23}Hz.
9) Calculate the frequency of violet light which has a time period of 1.33×10^{-15}s.
10) A wave has a frequency of 2MHz. Calculate the time period.

43. Refractive Index

If we assume that the speed of light in a vacuum is the same as in air (this is a good enough approximation at GCSE and IGCSE), then the refractive index, n, of a medium (a medium could be glass, Perspex, water or diamond for example) is defined by:

$$refractive\ index, n = \frac{speed\ of\ light\ in\ air\ (or\ vacuum)}{speed\ of\ light\ in\ medium}$$

Refractive index has no units.
Speed of light in air (or vacuum) is measured in metres per second, m/s.
Speed of light in medium is measured in metres per second, m/s.

This formula can be written as:

$$n = \frac{c_{air\ (or\ vacuum)}}{c_{medium}}$$

with **c** meaning the 'speed of light'.

The reason why refractive index has no units can be explained by looking at the formula. The 'speed of light in air (or vacuum)' measured in **m/s**, is divided by 'speed of light in medium' measured in **m/s**. looking at the units we are dividing **m/s** by **m/s**. These cancel out and therefore there is no unit.

The refractive index of a medium can also be calculated using the formula:

$$refractive\ index, n = \frac{sine\ of\ the\ angle\ of\ incidence}{sine\ of\ the\ angle\ of\ refraction}$$

Refractive index has no units.
Sine of the angle of incidence has no units.
Sine of the angle of refraction has no units.

It is important to know that when using this formula to calculate the refractive index of a medium (i.e. glass) that the ray of light must always be travelling from air (or a vacuum) to the medium. For example from air to glass if trying to calculate the refractive index of glass

This formula can be written as:

$$n = \frac{\sin i}{\sin r}$$

where **sin** means 'sine', **i** is 'the angle if incidence' and **r** is 'the angle of refraction'.

Note:
Both 'the angle of incidence' and the 'angle of refraction' are measured in degrees, °. However, '$\sin i$' and '$\sin r$' have no units!

To understand what we mean by 'angle of refraction', see figure 1 below:

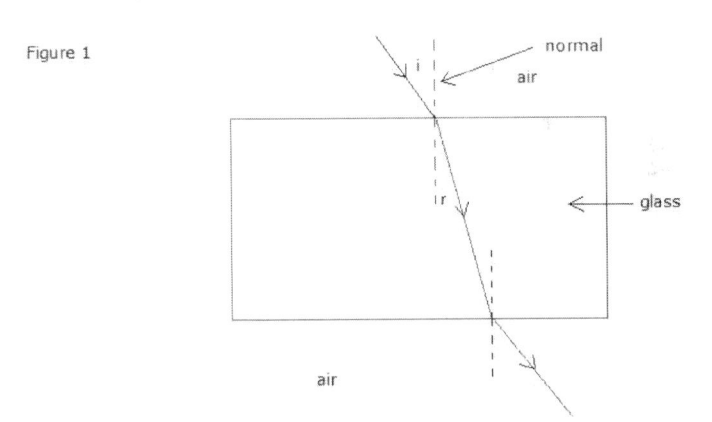

Figure 1

First of all I need to explain what the 'normal' is. **The 'normal' (the dashed line) is a reference line and all angles are measured from the normal. It is drawn where the ray of light hits the boundary between the 2 media, in this case air and glass, and is at 90° to the boundary.**
Figure 1 shows a ray of light entering a glass block from the air (the ray of light is shown by the line with the arrows on it). As the ray of light enters the glass block it changes direction - in this case it refracts 'towards the normal'. If we let **i**, the angle of incidence equal 20° and **r** equal 13°, can you see that the ray of light has 'bent' towards the normal. It was 20° from the normal and now it is only 13°. We use the word 'bent', but it is important to realise that in this context it means 'changes direction'.

When a ray of light passes from a medium of lower refractive index to a medium of higher refractive index (for example from air to glass) the light 'bends' or 'refracts' towards the normal (unless the ray of light is travelling

along the normal – see figure 2)
When a ray of light passes from a medium of higher refractive index to a medium of lower refractive index (for example from glass to air) the light 'bends' or 'refracts' away from the normal (unless the ray of light is travelling along the normal – see figure 2). This is shown in figure 1 as the light leaves the block.

When a ray of light is travelling along the normal as in figure 2, the path of the ray of light is unchanged.

Figure 2

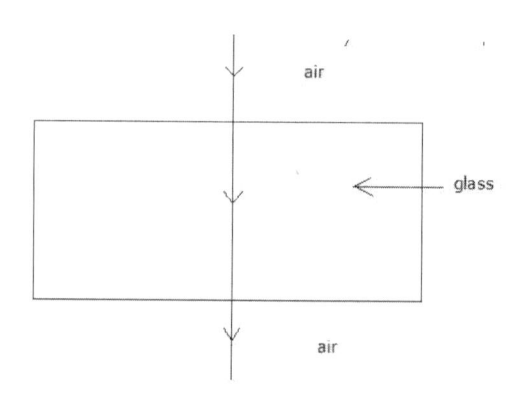

When a ray of light travels from a medium of lower refractive index to one of higher refractive index, for example from air to glass, the ray of light slows down. However, the frequency of the light does not change. If we go back to section 41, v = f x λ. Looking at this formula, if the wave speed, **v**, has decreased but the frequency has remained unchanged, the wavelength must have decreased. Let's use some simple numbers to explain this. If v = 10m/s, f = 5Hz and λ = 2m (from v = f x λ, 10 = 5 x 2). If **v** has now decreased to 5m/s and the frequency remained unchanged at 5Hz then the wavelength must have decreased to 1m (from v = f x λ, 5 = 5 x 1).
The opposite happens when light travels from a medium of higher refractive index to one of lower refractive index. This time the ray of light speeds up, but again the frequency remains unchanged, which means that the wavelength must increase.

If we take a look at rearranging the formula:

$$n = \frac{\sin i}{\sin r}$$

To make **sin i** the subject of the formula we need to multiply both sides by **sin r**. This gives:

$$n \times \sin r = \sin i$$

Therefore,
$$\sin i = n \times \sin r$$

I will explain in the examples how to find the angle of incidence, **i**, from **sin i**. If we go back to $n \times \sin r = \sin i$, if we want to make **sin r** the subject of the

Physics calculations for GCSE & IGCSE

formula, we need to divide both sides by **n**. This will give:

$$\sin r = \frac{\sin i}{n}$$

I will explain in the examples how to find the angle of refraction, **r**, from **sin r**.

In a similar way, we can rearrange the formula:

$$n = \frac{C_{air\ (or\ vacuum)}}{C_{medium}}$$

To make $C_{air\ (or\ vacuum)}$ the subject of the formula, we need to multiply both sides by C_{medium}. This gives:

$$n \times C_{medium} = C_{air\ (or\ vacuum)}$$

To make C_{medium} the subject of the formula from, $n \times C_{medium} = C_{air\ (or\ vacuum)}$, we need to divide both sides by **n**. This gives:

$$C_{medium} = \frac{C_{air\ (or\ vacuum)}}{n}$$

Example 1

A ray of light travelling from air to diamond has an angle of incidence of 60° and an angle of refraction of 21°. Calculate the refractive index of diamond.

i = 60°
r = 21°
n = ?

$$n = \frac{\sin i}{\sin r}$$

$$n = \frac{\sin 60}{\sin 21}$$

$$n = \frac{0.8660254038}{0.3583679495}$$

n = 2.4

Example 2

A ray of light is travelling from air to water. If the angle of incidence in air is 25.0°, calculate the angle of refraction if the refractive index of water is 1.33.

n_{water} = 1.33
i = 25.0°
r = ?

$$n = \frac{\sin i}{\sin r}$$

$$\sin r = \frac{\sin i}{n}$$

$$\sin r = \frac{\sin 25.0}{1.33}$$

$$\sin r = \frac{0.4226182617}{1.33}$$

$$\sin r = 0.3177580915$$

At the moment we have found the sine of the angle **r**. To find **r** using your calculator you need to 'inverse sine' this answer. On most calculators to do this you press SHIFT then SIN and then input the number (with some calculators you may have to input the number before you press SHIFT then SIN). 'Inverse sine' can be represented by **sin⁻¹**.

Therefore,

$$r = \sin^{-1}(0.3177580915)$$

$$r = 18.5°$$

Example 3

A ray of light is travelling from air to glass. If the angle of refraction in the glass is 33.0° and the refractive index of glass is 1.50, calculate the angle of incidence in air.

i = ?
r = 33.0°
n_{glass} = 1.50

$$n = \frac{\sin i}{\sin r}$$

$$\sin i = n \times \sin r$$

$$\sin i = 1.50 \times \sin 33.0$$

$$\sin i = 1.50 \times 0.544639035$$

$$\sin i = 0.8169585525$$

$$i = \sin^{-1}(0.8169585525)$$

$$i = 54.8°$$

Example 4

If the speed of light in air is 300,000,000m/s and the refractive index of glass is 1.50, calculate the speed of light in glass.

C_{air} = 300,000,000m/s
n_{glass} = 1.50
C_{glass} = ?

$$n = \frac{C_{air \text{ (or vacuum)}}}{C_{glass}}$$

$$C_{glass} = \frac{C_{air \text{ (or vacuum)}}}{n}$$

$$C_{glass} = \frac{300,000,000}{1.50}$$

$$C_{glass} = 200,000,000 \text{ m/s}$$

Questions

1) A ray of light travelling from air to water has an angle of incidence of 40.0° and an angle of refraction of 28.9°. Calculate the refractive index of water.
2) A ray of light is travelling from air to glass. If the angle of refraction in the glass is 40.0° and the refractive index of glass is 1.50, calculate the angle of incidence in air.
3) A ray of light is travelling from air to water. If the angle of incidence in air is 35.0°, calculate the angle of refraction if the refractive index of water is 1.33.
4) A ray of light travelling from air to ice has an angle of incidence of 45.0° and an angle of refraction of 32.7°. Calculate the refractive index of ice.
5) If the speed of light in air is 3.0×10^8 m/s and the refractive index of diamond is 2.4, calculate the speed of light in diamond.
6) If the speed of light in oil is 2.34×10^8 m/s and the speed of light in air is 3.00×10^8 m/s, calculate the refractive index of oil.
7) A ray of light is travelling from air to glass. If the angle of refraction in the glass is 25.0° and the refractive index of glass is 1.50, calculate the angle of incidence in air.
8) A ray of light is travelling from air to water. If the angle of incidence in air is 70.0°, calculate the angle of refraction if the refractive index of water is 1.33.

44. Total internal reflection and critical angle

Total internal reflection can be explained using figures 1, 2 and 3:

Figure 1

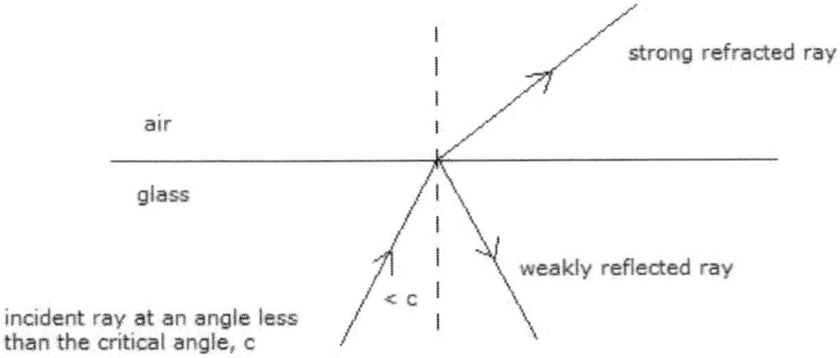

strong refracted ray

air

glass

weakly reflected ray

incident ray at an angle less than the critical angle, c

< c

When light passes from a medium of high refractive index to low refractive, for example glass to air, if the incident ray is less than the critical angle (if the angle of refraction is 90° then the incident ray is at the critical angle), then as

in figure 1, there is a weakly (not very bright) reflected ray at the same angle as the incident ray and a strongly (bright) refracted ray.
If the angle of incidence of the light is increased, the reflected ray becomes stronger (again this will be at the same angle as the angle of incidence) and the angle of refraction increases while also becoming weaker (less bright).

Figure 2

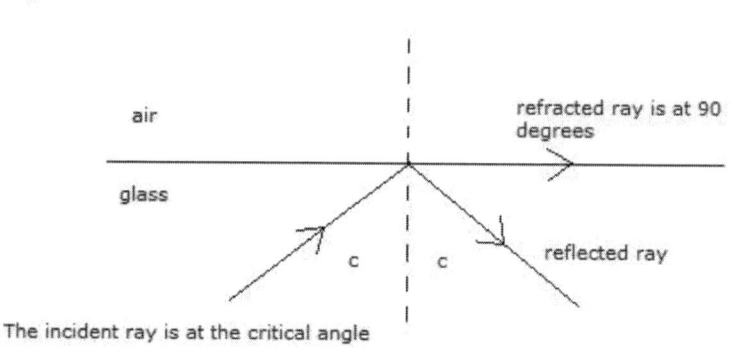

refracted ray is at 90 degrees

air

glass

reflected ray

c   c

The incident ray is at the critical angle

When the angle of incidence has increased so that the angle of refraction is 90°, the incident ray is said to be at the critical angle. This is shown in figure 2 above.

If the incident ray is increased so that it is now **greater** than the critical angle, **all** of the light is now reflected at the glass-air boundary and none is refracted. This is known as total internal reflection. See figure 3 below.

Figure 3

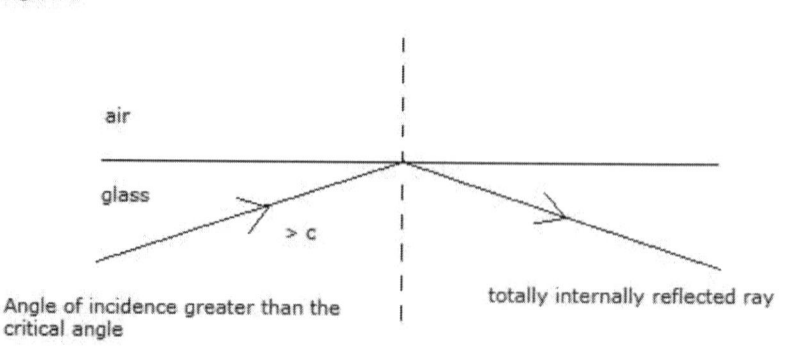

air

glass

> c

totally internally reflected ray

Angle of incidence greater than the critical angle

There are 2 conditions for total internal reflection:
1) The ray of light must be travelling from a medium of higher refractive index to a medium of lower refractive index.
2) The angle of incidence must be greater than the critical angle.

The critical angle, **c**, is different for different mediums and can be calculated using the formula below:

$$\sin c = \frac{1}{n}$$

Where **c** is the critical angle measured in degrees and **n** is the refractive index of the medium that the ray of light is incident in. **This formula can only be used if the medium of lower refractive index is air (or a vacuum if we assume that the refractive index of a vacuum is the same as that of air).**

Note:
As with section 43 where '$\sin i$' and '$\sin r$' had no units, this is also the case for '$\sin c$'. Remember also that refractive index, **n** has no units!

To rearrange the formula to make **n** the subject, we firstly need to multiply both sides by **n**. This gives:

$$n \times \sin c = 1$$

If we now divide both sides by **sin c**, this will give:

$$n = \frac{1}{\sin c}$$

Example 1

The refractive index of diamond is 2.42. Calculate the critical angle of diamond.

n = 2.42
c = ?

$$\sin c = \frac{1}{n}$$

$$\sin c = \frac{1}{2.42}$$

$$\sin c = 0.4132231405$$

At the moment we have found the sine of the critical angle, **c**. To find **c** using your calculator you need to 'inverse sine' this answer. On most calculators to do this you press SHIFT then SIN and then input the number (with some calculators you may have to input the number before you press SHIFT then SIN).
'Inverse sine' can be represented by **sin⁻¹**.
Therefore,

$$c = \sin^{-1}(0.4132231405)$$

$$c = 24.4°$$

Example 2

The critical angle for glass is 41.8°. Calculate the refractive index of glass.

c = 41.8°
n = ?

$$\sin c = \frac{1}{n}$$

$$n = \frac{1}{\sin c}$$

$$n = \frac{1}{\sin 41.8}$$

$$n = \frac{1}{0.6665324702}$$

$$n = 1.50$$

Questions

1) The refractive index of ice is 1.31. Calculate the critical angle of ice.
2) The critical angle for oil is 51.4°. Calculate the refractive index of oil.
3) The refractive index of Ruby is 1.76. Calculate the critical angle of Ruby.
4) The critical angle for Quartz is 40.5°. Calculate the refractive index of Quartz.
5) The critical angle for water is 48.8°. Calculate the refractive index of water.
6) The refractive index of Ethanol is 1.36. Calculate the critical angle of Ethanol.
7) The critical angle for paraffin is 44.0°. Calculate the refractive index of paraffin.

45. P = 1/f

Before looking at the formula for the power of a lens, we need to look at some of the basic properties of lenses. Lenses can either be classed as convex (also known as converging) or concave (also known as diverging). Some examples of these types of lenses are shown in figures 1 and 2:

Figure 1 - Some examples of convex lenses

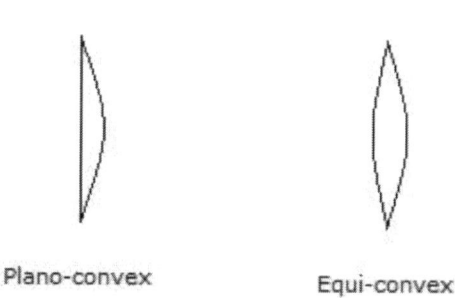

Plano-convex   Equi-convex

If you look at the examples in figure 1 you can see that convex (or converging) lenses are thickest in the centre of the lens. In figure 2 you can see that concave (or diverging) lenses are thinnest at the centre of the lens.

Figure 2 - Some examples of concave lenses

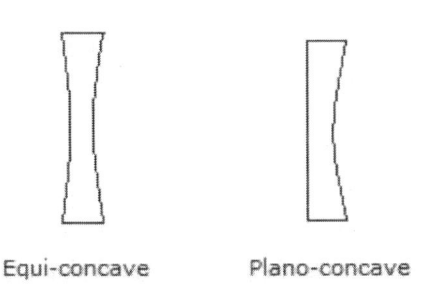

Equi-concave   Plano-concave

To help explain convex lenses see figure 3 below:

As the rays of light pass through the lens they are refracted (except the one that passes along the principal axis) and converge to a point. This is why they are also known as converging lenses. The rays of light converge at a point known as the focal point or principal focus.
The focal point of a lens is defined as the point to which rays of light parallel and close to the principal axis converge or from which they appear to diverge. In figure 3 below they converge.
The line which passes through the centre of the lens at a right angle to it is known as the principal axis.

Figure 3 - Action of a convex lens   convex lens symbol

← principal axis

Rays of light parallel and close to the principal axis

Focal point or principal focus

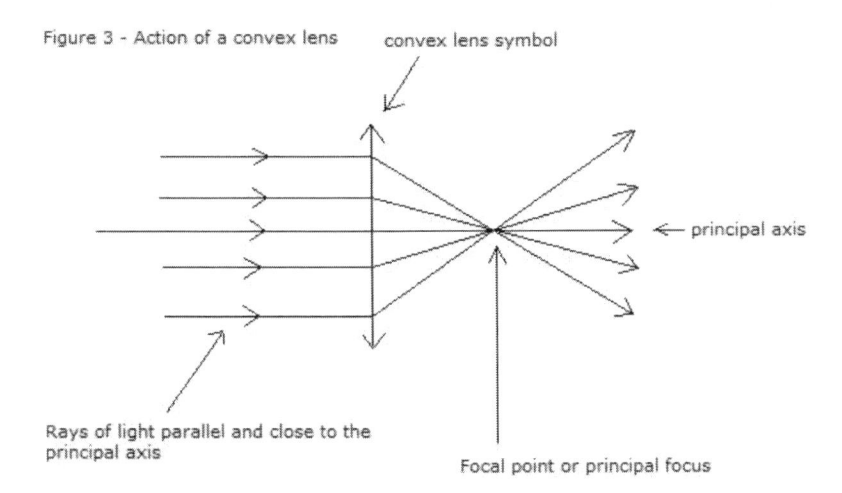

To help explain concave lenses see figure 4:

As the rays of light pass through the lens they are refracted (except the one that passes along the principal axis) and diverge away from the principal axis. This is why they are also known as diverging lenses.

With the concave lens in figure 4 the rays of light appear to have come from the focal point.

This is shown on the diagram by the dashed lines. These are not real rays of light, they are virtual rays of light. This is where our eye thinks that the light is coming from and the image that forms here is known as a virtual image.

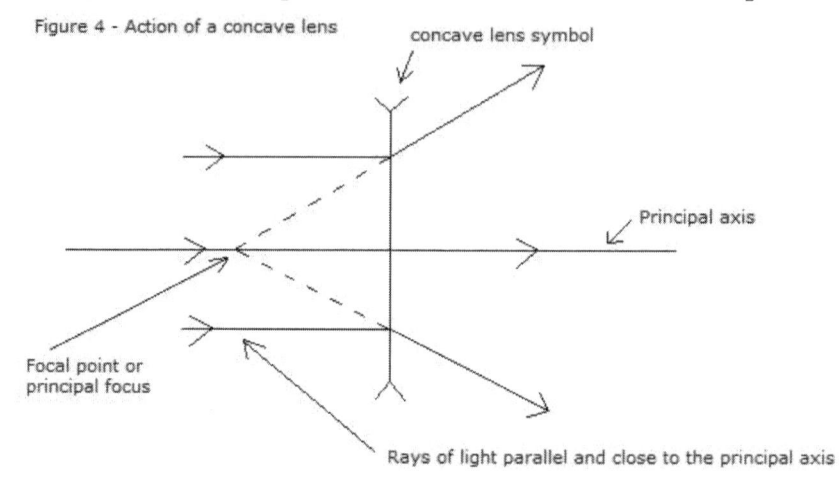

Figure 4 - Action of a concave lens

concave lens symbol

Principal axis

Focal point or principal focus

Rays of light parallel and close to the principal axis

A virtual image is one that cannot be formed on a screen and rays of light only appear to come from the image. All images formed by concave lenses are virtual.

If we take a look again at our definition of focal point, **the focal point of a lens is defined as the point to which rays of light parallel and close to the principal axis converge or from which they appear to diverge.** In figure 4 above they diverge.

Going back to figure 3, an image formed at the focal point here has been formed from real rays of light. **A real image is one through which actual rays of light pass and therefore the image can be formed on a screen.** With convex lenses most images are nearly always real.

The power of a lens can be calculated using the formula below:

$$Power = \frac{1}{focal\ length}$$

The unit of measurement for the power of a lens is the dioptre, D.
The focal length is measured in metres, m.

The focal length of a lens is the distance from the centre of the lens along the principal axis to the focal point. Light can enter a lens from both sides so a lens will have 2 focal points, one on each side of the lens. They will both be the same distance from the centre of the lens.

This formula can be written as:

$$P = \frac{1}{f}$$

If we want to make **f** the subject of the formula, the first thing we need to do is multiply both sides by **f**. This gives:

$$P \times f = 1$$

To now make **f** the subject of the formula, we need to divide both sides by **P**. This will give:

$$f = \frac{1}{P}$$

It is important to note the following points with this formula:
1) The power of a convex (converging) lens is positive. Looking at the formula this must also mean that the focal length of a convex (converging) lens is also positive.
2) The power of a concave (diverging) lens is negative. Looking at the formula this must also mean that the focal length of a concave (diverging) lens is also negative.

Going back to the power of a lens formula:

$$P = \frac{1}{f}$$

the smaller **f** is, the greater the power of the lens (if f = 0.2m, P = 5D, if f = 0.1m, P = 10D). This means that more powerful lenses refract the light more which is why their focal length, **f** is smaller.

It is also useful to know that for a given focal length, a lens made from a material of high refractive index will be thinner than a material with a lower refractive index. This is because if you have two lenses identical in shape but made from different materials, the lens with the higher refractive index refracts the light more and therefore has a shorter focal length.

Example 1

A converging lens has a focal length of 5cm. Calculate the power of the lens.

A converging lens has a positive focal length so f = +5cm. However, the focal length must be in metres. To convert centimetres (cm) to metres (m) you divide by 100.
Alternatively, from section 2, 5cm = 5 x 10^{-2}m = 0.05m **or** 5cm = 5 x 0.01 = 0.05m.

f = +5cm = +0.05m
P = ?

$$P = \frac{1}{f}$$

$$P = \frac{1}{0.05}$$

P = 20D

Example 2

A diverging lens has a power of -10D. Calculate the focal length.

P = -10D
f = ?

$$P = \frac{1}{f}$$

$$f = \frac{1}{P}$$

$$f = \frac{1}{-10}$$

f = -0.1m

Example 3

A diverging lens has a focal length of -5mm. Calculate the power of the lens.

From section 2, 5mm = 5 x 10^{-3}m = 0.005m **or** 5mm = 5 x 0.001 = 0.005m. Alternatively, to convert **mm** to **m** you divide by 1000 (5/1000 = 0.005). Therefore:

f = -0.005m
P = ?

$$P = \frac{1}{f}$$

$$P = \frac{1}{-0.005}$$

P = -200D

Questions

1) Calculate the power of a converging lens if the focal length is:
   a) 1.0m b) 0.5m c) 0.25m d) 10mm e) 80.0cm
2) Calculate the power of a diverging lens if the focal length is:
   a) -0.05m b) -0.5m c) -200mm d) -0.600m e) -15.0mm
3) Calculate the focal length of a lens if the power is:
   a) 8.00D b) -4.0D c) 15D d) 100D e) -50D
4) Calculate the power of a converging lens if the focal length is 7.00mm.
5) Calculate the power of a diverging lens if the focal length is -30cm.
6) Calculate the focal length of a diverging lens if the power is -40D.
7) Calculate the focal length of a convex lens if the power is +45D.

46. 1/u + 1/v = 1/f

The lens formula is given by:

$$\frac{1}{u} + \frac{1}{v} = \frac{1}{f}$$

u = object distance (this is the distance of the object, for example a tree or person, from the lens).
v = image distance (this is the distance of the image, for example of a tree or person, from the lens).
f = focal length of the lens.
This can be shown in figure 1 below:

Figure 1

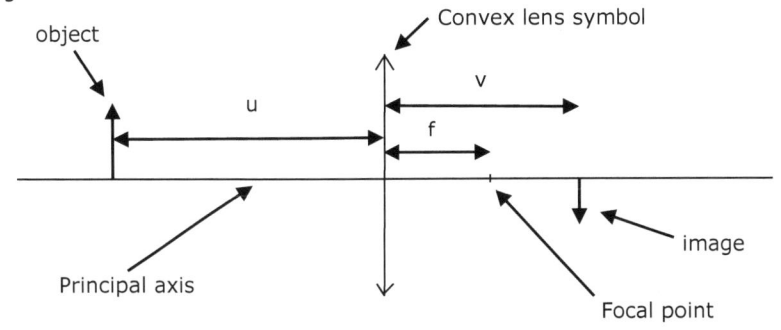

The distances **u** and **v** and the focal length **f** are all measured from the centre of the lens along the principal axis as shown in figure 1.
With the lens formula, **u**, **v** and **f** can be measured in any unit of length as long as the same unit is used for all of them. You could therefore measure them all in millimeters, or all of them in centimetres for example.
The above formula also requires a sign convention. We are going to use the **real is positive** sign convention. Using this convention means that:
 1) If the lens is convex (converging), **f** is positive.
 2) If the lens is concave (diverging), **f** is negative.
 3) If **v** is positive, the image is real.
 4) If **v** is negative, the image is virtual.
 5) If **u** is positive, the object is real.
 6) If **u** is negative, the object is virtual.

If we take a look at rearranging the formula, if we want to make **1/u** the subject of the formula we need to subtract **1/v** from both sides (remember that this is because **1/v** is added to **1/u**. To remove **1/v** from the left hand side of the formula we do the 'opposite' or 'inverse' which means that we have to subtract it). This gives:

$$\frac{1}{u} = \frac{1}{f} - \frac{1}{v}$$

I will show you how to calculate **u** from **1/u** when I do the examples.
Going back to the original formula:

$$\frac{1}{u} + \frac{1}{v} = \frac{1}{f}$$

If we want to make **1/v** the subject of the formula we need to subtract **1/u** from both sides. This gives:

$$\frac{1}{v} = \frac{1}{f} - \frac{1}{u}$$

Again, I will show you how to calculate **v** from **1/v** when I do the examples.

Example 1

A converging lens has a focal length of 10cm. If the object is placed 30cm away from the lens, calculate the image distance, v.

The lens is a converging one so f = +10cm.
The object is real so u = +30cm (to be able to place an object a certain distance from a lens the object must be real. This is because you have moved it and can therefore touch it).
v = ?

$$\frac{1}{u} + \frac{1}{v} = \frac{1}{f}$$

$$\frac{1}{v} = \frac{1}{f} - \frac{1}{u}$$

$$\frac{1}{v} = \frac{1}{10} - \frac{1}{30}$$

You can use a calculator for this, but it is not difficult to work out. To subtract fractions, both denominators need to be the same. Using equivalent fractions, 1/10 is equal to 3/30 (here we have multiplied both the numerator and denominator by 3), therefore, replacing 1/10 with 3/30 we get:

$$\frac{1}{v} = \frac{3}{30} - \frac{1}{30} = \frac{2}{30} = \frac{1}{15}$$

If you look at the fraction 2/30, if we divide both numerator and denominator by 2, we get 1/15. Therefore:

$$\frac{1}{v} = \frac{1}{15}$$

To make **v** the subject of the formula, multiplying both sides by **v** gives:

$$1 = \frac{1 \times v}{15}$$

This is the same as:

$$1 = \frac{v}{15}$$

If we now multiply both sides by 15 we get:

$$1 \times 15 = v$$

$$15 = v$$

Therefore,
$$v = 15\text{cm}$$

This is a positive value and means that we have a real image.

Example 2

A lens has an object placed 10cm from the lens. If the image distance is -20cm, calculate the focal length of the lens.

The object is real so u = +10cm.
v = -20cm (the image must be virtual because **v** is negative).
Remember that a virtual image is one that cannot be formed on a screen and rays of light only appear to come from the image.

f = ?

$$\frac{1}{u} + \frac{1}{v} = \frac{1}{f}$$

$$\frac{1}{f} = \frac{1}{u} + \frac{1}{v}$$

$$\frac{1}{f} = \frac{1}{10} + \frac{1}{-20}$$

(remembering that '+- = -', this equals:

$$\frac{1}{f} = \frac{1}{10} - \frac{1}{20}$$

Again, you can use a calculator for this but it is not difficult to do without a calculator. To subtract fractions, both denominators need to be the same. Using equivalent fractions, 1/10 is equal to 2/20 (here we have multiplied both the numerator and denominator by 2), therefore, replacing 1/10 with 2/20 we get:

$$\frac{1}{f} = \frac{2}{20} - \frac{1}{20}$$

$$\frac{1}{f} = \frac{1}{20}$$

To make **f** the subject of the formula, we just follow the same method as example 1. Multiplying both sides by **f** gives:

$$1 = \frac{1 \times f}{20}$$

This is the same as:

$$1 = \frac{f}{20}$$

Multiplying both sides by 20 gives:

$$1 \times 20 = f$$

$$20 = f$$

Therefore,

$$f = 20\text{cm}$$

The focal length is positive, so we have a converging lens of focal length 20cm.

Example 3

A diverging lens has a focal length of -30cm. If a virtual image is produced 15cm from the lens, calculate the object distance, u.

f = -30cm
v = -15cm (the image is virtual so the image distance is negative)
u = ?

$$\frac{1}{u} + \frac{1}{v} = \frac{1}{f}$$

$$\frac{1}{u} = \frac{1}{f} - \frac{1}{v}$$

$$\frac{1}{u} = \frac{1}{-30} - \frac{1}{-15}$$

(remembering that '-- = +', this equals:

$$\frac{1}{u} = \frac{1}{-30} + \frac{1}{15}$$

Again, you can use a calculator for this but I will show you how to do it without a calculator. To add fractions, both denominators need to be the same. Using equivalent fractions, 1/15 is equal to 2/30 (here we have multiplied both the numerator and denominator by 2), therefore, replacing 1/15 with 2/30 we get:

$$\frac{1}{u} = \frac{1}{-30} + \frac{2}{30}$$

$$\frac{1}{u} = \frac{1}{30}$$

This can be rearranged in exactly the same way as in examples 1 & 2 to give:

u = 30cm.

This means that the object is real and 30cm from the lens.

Questions

1) A converging lens has a focal length of 15cm. If the object is placed 40cm away from the lens, calculate the image distance, v.
2) A lens has an object placed 10cm from the lens. If the image distance is -30cm, calculate the focal length of the lens.
3) A diverging lens has a focal length of -10.0cm. If a virtual image is produced 25.0cm from the lens, calculate the object distance, u.
4) A converging lens has a focal length of 6mm. If the object is placed 12mm away from the lens, calculate the image distance, v.
5) A converging lens has a focal length of 8cm. If the object is placed 12cm away from the lens, calculate the image distance, v.
6) A converging lens has a focal length of 15mm. If an image is produced 20mm from the lens, calculate the object distance, u.
7) A diverging lens has a focal length of -30cm. If a virtual image is

produced 20cm from the lens, calculate the object distance, u.

47. Magnification

The magnification produced by a lens can be calculated using the formula below:

$$magnification = \frac{image\ height}{object\ height}$$

The image height and object height can be measured in any unit, but when using the formula both heights must be in the same unit.
Magnification has no units.
The reason why magnification has no units can be explained by looking at the formula. For example, if the image height is measured in metres, **m** and this is divided by the object height measured also in metres, **m** then looking at the units we are dividing **m** by **m**. These cancel out and therefore there is no unit.

If the magnification is equal to 1, then 'image height = object height'.
If the magnification is less than 1, then 'image height < object height'.
If the magnification is greater than 1, then 'image height > object height'.
Remember that '<' means 'less than' and '>' means 'greater than'.

Taking a look at rearranging the formula, if we want to make **image height** the subject of the formula we need to multiply both sides by **object height**. This gives:

$$object\ height\ x\ magnification = image\ height$$

Therefore,

$$image\ height = object\ height\ x\ magnification$$

To make object height the subject of the formula, if we go back to *'object height x magnification = image height'*, if we divide both sides by **magnification** this gives:

$$object\ height = \frac{image\ height}{magnification}$$

Example 1

An object of height 5.0cm has an image produced by a lens of height 2.5cm. Calculate the magnification of the lens.

object height = 5.0cm
image height = 2.5cm
magnification = ?

$$magnification = \frac{image\ height}{object\ height}$$

$$magnification = \frac{2.5}{5.0}$$

magnification = 0.50

Example 2

An object of height 20mm has an image produced by a lens of height 6cm. Calculate the magnification of the lens.

Firstly we need both units of height to be the same. I will convert the 20mm into centimetres, cm. There are 10mm in 1cm, therefore in order to convert **mm** to **cm** we need to divide by 10.

$$\frac{20}{10} = 2$$

object height = 20mm = 2cm
image height = 6cm
magnification = ?

$$magnification = \frac{image\ height}{object\ height}$$

$$magnification = \frac{6}{2}$$

$$magnification = 3$$

Example 3

A lens produces a magnification of 0.25. If the image height is 1mm, calculate the object height.

magnification = 0.25
image height = 1mm
object height = ?

$$magnification = \frac{image\ height}{object\ height}$$

$$object\ height = \frac{image\ height}{magnification}$$

$$object\ height = \frac{1}{0.25}$$

$$object\ height = 4mm$$

Remember that both units must be the same. The image height was given in **mm** so the object height must also be measured in **mm**.

Example 4

A lens produces a magnification of 2. If the object height is 10cm, calculate the image height.

magnification = 2
object height = 10cm
image height = ?

$$magnification = \frac{image\ height}{object\ height}$$

$$image\ height = object\ height \times magnification$$

$$image\ height = 10 \times 2$$

image height = 20cm

Questions

1) A lens produces a magnification of 4. Calculate the image height if the object height is 2cm.
2) A lens produces a magnification of 0.5. If the image height is 8mm, calculate the object height.
3) An object of height 30cm has an image produced by a lens of height 15cm. Calculate the magnification of the lens.
4) An object of height 40cm has an image produced by a lens of height 10cm. Calculate the magnification of the lens.
5) A lens produces a magnification of 0.8. If the image height is 16mm, calculate the object height.
6) A lens produces a magnification of 5. Calculate the image height if the object height is 4cm.
7) An object of height 200mm has an image produced by a lens of height 2cm. Calculate the magnification of the lens.

48. Half-life

Before commencing this topic there are some terms that we need to define:

Mass number – This number gives the number of protons and neutrons in an atom. Protons and neutrons are found in the nucleus of the atom.

Proton (or atomic) number – This is the number of protons in the nucleus of an atom. The proton number defines the element. For example if the proton number of an element is 1, the element is hydrogen. If the proton number of an element is 8, the element is oxygen.

Nuclide – An atom that is characterized by its mass number and proton number (at a higher level of Physics we would also need to include energy state). If any of these change, the atom becomes a different nuclide.

Alpha particle – A Helium nucleus. This comprises 2 protons and 2 neutrons.

Alpha radiation – A stream of alpha particles.

Beta particle – A fast moving electron (at a higher level the positron which is the antiparticle of the electron is also classed as a beta particle).

Beta radiation – A stream of beta particles.

Gamma radiation – Very high frequency (or very short wavelength) electromagnetic radiation.

Radioactivity – The spontaneous decay of the nuclei of some nuclides that involves the emission of alpha or beta radiation and sometimes gamma radiation.

Radionuclide – A nuclide that is radioactive.

Isotope – Two or more nuclides that have the same proton (atomic) number but different mass number. For example there are 3 isotopes of hydrogen as shown below:

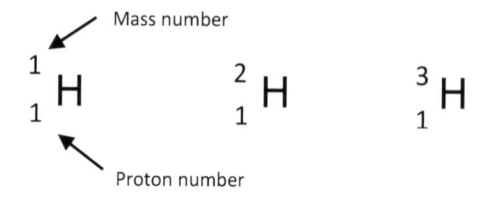

$$^{1}_{1}H \qquad ^{2}_{1}H \qquad ^{3}_{1}H$$

Mass number (top), Proton number (bottom)

Radioisotope – A radioactive isotope.

When a radionuclide decays it emits alpha or beta radiation. If the nucleus has any extra (or surplus) energy this will be given off as gamma radiation. Radioactivity is a totally random process and it is not possible to predict when a nucleus of a radionuclide will decay. However it is possible to calculate the time it takes, on average, for half of the nuclei in a sample of a radionuclide to decay. This is known as the half-life.

There are two definitions for half-life:
1) Half-life is the **time** taken, on average, for half of the nuclei in a sample of a radionuclide to decay.
2) Half-life is the **time** taken for the count rate of a sample of a radionuclide (or a radioisotope) to fall to half of its initial value.

It is important to remember that half-life is a time! Half-lives of radionuclides can vary from fractions of a second to billions of years.

Example 1

A radionuclide has an initial count rate of 400 counts per second (c/s). If the half-life of the radionuclide is 10 days, calculate the count rate after 40 days.

The first thing to calculate is how many half-lives there are in 40 days. This is done as follows:

$$\frac{40}{10} = 4$$

When solving problems like this it is best to use the method shown below:

400 → 200 → 100 → 50 → 25

Each arrow represents a half-life. All you need to do is start at the initial count rate and keep halving until you have the required number of half-lives. You also need to remember that each half-life is represented by an arrow. In this example there are 4 half-lives, so we need 4 arrows.

Therefore after 40 days the count rate will have reduced to 25c/s.

Example 2

Carbon-14 has a half-life of 5700 years (The 14 represents the mass number of the radionuclide. It is common for radionuclides or radioisotopes to be represented in this way, with the name of the radionuclide followed by the mass number). Calculate the fraction of carbon-14 that remains after 17,100 years.

Again, the first thing that we need to do is calculate the number of half-lives of carbon-14 in 17,100 years. This is done as follows:

$$\frac{17,100}{5700} = 3$$

We do not have a count rate in this question so we have to approach it slightly differently. We need to take the whole sample of carbon-14 as being equal to 1. We then keep halving for the required number of half-lives as shown below:

1 ⟶ ½ ⟶ ¼ ⟶ 1/8

Remembering that each arrow represents a half-life, all you have to do with this type of question is start with the number 1 and keep halving for the required number of half-lives (in this example, 3).

Therefore, the fraction of carbon-14 that remains in the sample after 17,100 years is 1/8 (one eighth).
An interesting point to note here is that carbon-14 decays into nitrogen-14 which is stable (not radioactive). **What this means however is that if we have a sample of carbon-14, 17,100 years later 1/8 of this sample will be carbon-14 while 7/8 (seven eighths) of the sample will be nitrogen-14!**

Example 3

Calculate the half-life of a radionuclide from the graph below:

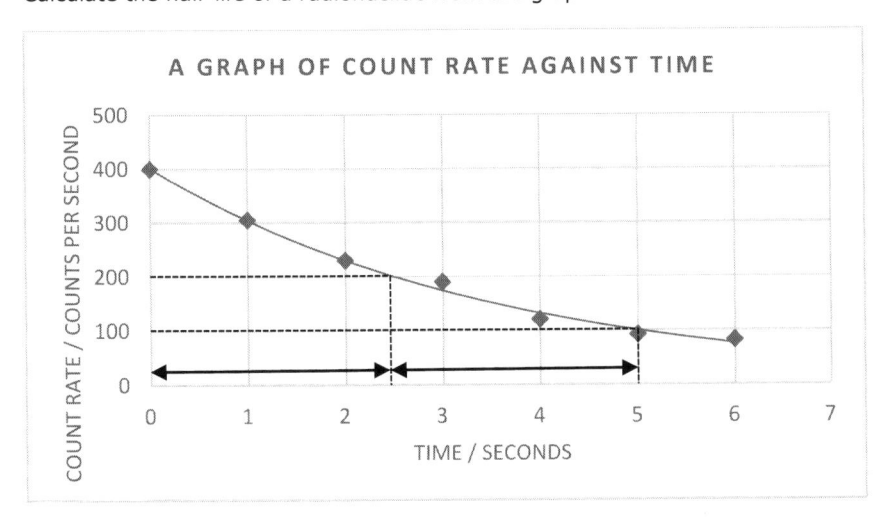

A GRAPH OF COUNT RATE AGAINST TIME

(Graph: Count rate / counts per second vs Time / seconds, showing exponential decay from 400 at t=0, ~300 at t=1, ~225 at t=2, ~190 at t=3, ~125 at t=4, ~100 at t=5, ~90 at t=6)

The half-life can be calculated from the graph using the following steps:
1) From the graph start with the initial count rate and half it. In this example the initial count rate is 400c/s. Half of this is 200c/s.
2) As shown in the graph, read horizontally from this value to the line of best fit. Once the line of best fit is reached, read vertically down to the time axis. This will then give the half-life (2.5 seconds in this example).
3) It is better to get another value and take an average. If we start from 200c/s, half of this will be 100c/s. As shown in the graph above, 200c/s occurs at 2.5 seconds and 100c/s occurs at 5 seconds. Therefore the time for the count rate to go from 200c/s to 100c/s will equal:

$$5 - 2.5 = 2.5 \text{ seconds}$$

4) Finally, using the 2 values calculated for the half-life we can calculate a mean (or average). This is calculated by adding up the values and dividing by the number of values that you have. Therefore the mean value for the half-life will be:

$$\text{mean half-life} = (2.5 + 2.5) \div 2$$

$$\text{mean half-life} = 5 \div 2$$

$$\text{mean half-life} = 2.5 \text{ seconds}$$

Questions

1) A radionuclide has an initial count rate of 1000 counts per second (c/s). If the half-life of the radionuclide is 19 minutes, calculate the count rate after 57 minutes.
2) Americium-242 has a half-life of 16 hours. Calculate the fraction of Americium-242 that remains after 64 hours.
3) Cobalt-60 has a half-life of 5.3 years. Calculate the fraction of Cobalt-60 that remains after 26.5 years.
4) A radionuclide has an initial count rate of 600c/s. If the half-life of the radionuclide is 15 days, calculate the count rate after 45 days.
5) A radionuclide has an initial count rate of 800c/s. If the half-life of the radionuclide is 5 minutes, calculate the count rate after 25 minutes.
6) Iodine-131 has a half-life of 8 days. Calculate the fraction of Iodine-131 that remains after 48 days.
7)

Time/minutes	Count rate/counts per second
0	960
2	480
4	230
6	120
8	60
10	30

The data in the table above was from a radionuclide.
a) Plot a graph of count rate against time (Hint: Count rate is on the y-axis, time is on the x-axis).
b) From the graph calculate the half-life of the radionuclide.

49. Graphs

Consider the distance-time graph of an object as shown in figure 1 below:

In order to calculate the speed from a distance-time graph like the one shown in figure 1, you need to calculate the gradient of the line. The gradient of a line can be calculated as shown:

$$gradient = \frac{change\ in\ y\ values}{change\ in\ x\ values}$$

This can be written as:

$$gradient = \frac{\Delta y}{\Delta x}$$

where **Δ**, is the Greek letter 'delta' which means 'change in'. **Δy** and **Δx** can be represented as shown in figure 1 below.

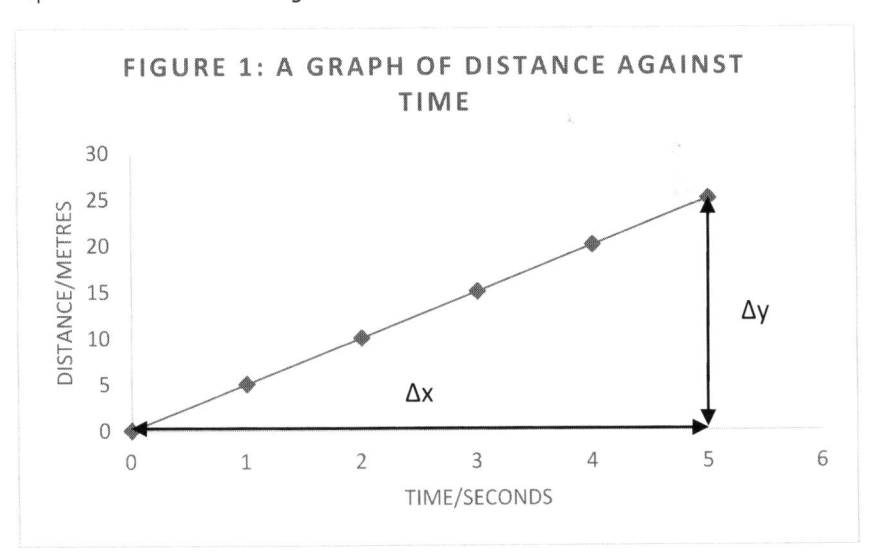

FIGURE 1: A GRAPH OF DISTANCE AGAINST TIME

In the graph above, **Δy = 25m** and **Δx = 5s**. Therefore the gradient of this graph is:

$$gradient = \frac{\Delta y}{\Delta x}$$

$$gradient = \frac{25m}{5s}$$

gradient = 5m/s.

In this example, from the gradient formula, we have divided a distance of 25m by a time of 5s. Looking at the units, we have divided **metres** by **seconds**. This is why the gradient of a distance-time graph is equal to the speed. The units for the gradient are **m/s** which are the same as the units for speed.

Therefore, for a distance-time graph:

The gradient of a distance-time graph is equal to the speed.

Looking at the graph in figure 1 you can see that the gradient (the steepness of the line) does not change, it is constant. If the gradient is the same, this means that the speed stays the same. **Therefore a graph like the one in figure 1 where the gradient does not change shows that the object is travelling at a constant speed.**

Also, if the gradient of a distance-time graph is equal to the speed we can also conclude the following:

The greater the gradient (the steeper the line) the greater the speed.

If we now consider the distance-time graph of an object as shown in figure 2 below. If you look at the graph you can see that it is curving upwards. As it curves upwards it gets steeper. If the line is steeper, this means that the gradient is greater!

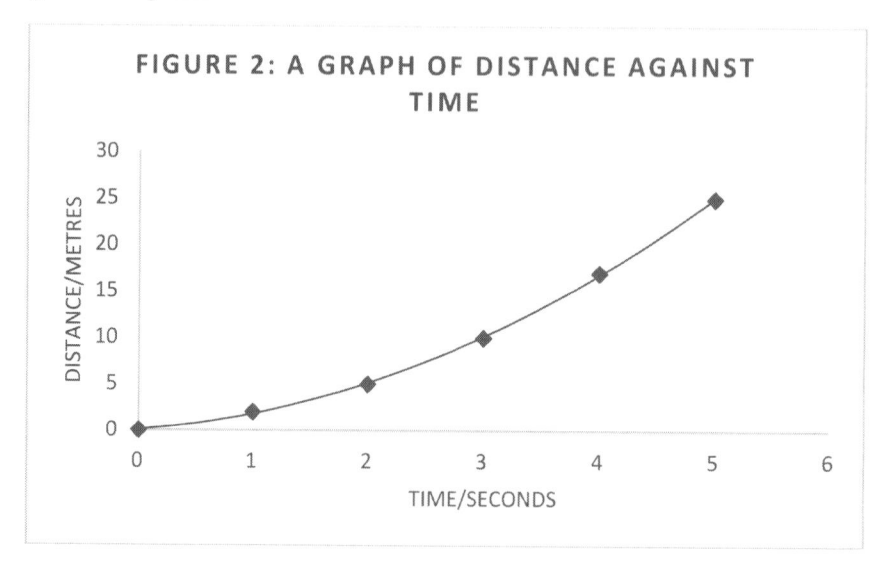

FIGURE 2: A GRAPH OF DISTANCE AGAINST TIME

If the gradient is greater, this means that the speed must be greater. If the speed is greater this means that the object must be accelerating! With a distance–time graph, if the graph is curving upwards as shown in figure 2, this means that the object is accelerating.
In order to calculate the speed at an instant in time from a graph like that shown in figure 2, a tangent needs to be drawn to the curve at that point (**the tangent to the curve at a point is a straight line that touches the curve**

at that point without crossing it). The gradient of the tangent is equal to the gradient of the curve at that point.

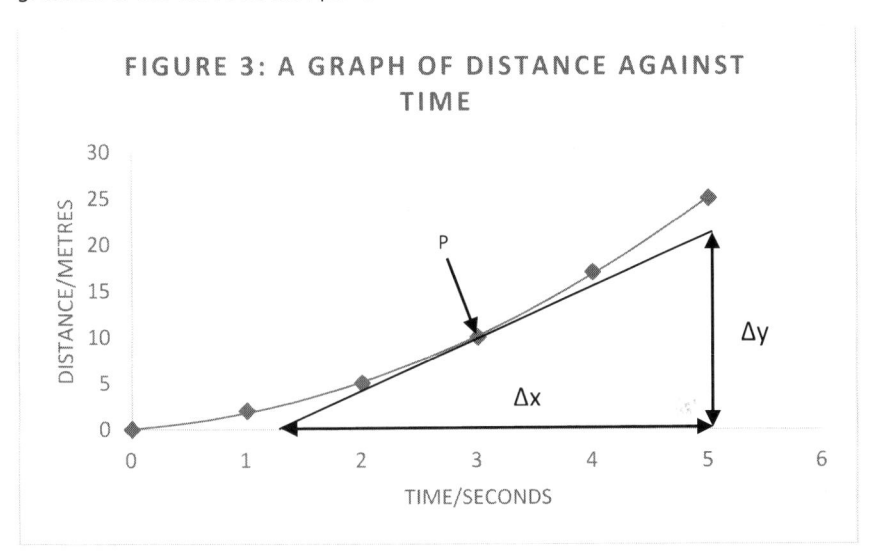

FIGURE 3: A GRAPH OF DISTANCE AGAINST TIME

[Graph showing distance (metres) on y-axis from 0 to 30, against time (seconds) on x-axis from 0 to 6. A curve passes through points with a tangent drawn at point P, showing Δx and Δy as the horizontal and vertical changes of the tangent triangle.]

If we consider the point, **P**, in figure 3 above, in order to calculate the speed at point **P**, you need to draw a tangent to the curve at point **P** as shown. The gradient of the tangent will give you the gradient of the curve at that point and the gradient of the curve at that point gives the speed at that point. The gradient at point **P** can be calculated using the gradient formula.

$$gradient = \frac{\Delta y}{\Delta x}$$

$$gradient\ at\ point\ P = \frac{21 - 0}{5 - 1.3}$$

$$gradient\ at\ point\ P = \frac{21}{3.7}$$

gradient at point P = 5.7m/s

Therefore the speed at point P = 5.7m/s.

Figure 4 shows a sketch of a distance-time graph. Line A has a gradient equal to zero. This is because Δy = 0 (gradient = Δy/Δx, therefore if Δy is zero the gradient will be zero). For example, if Δy = 0m and Δx = 5s then the gradient, Δy/Δx = 0/5 = 0m/s.

Figure 4

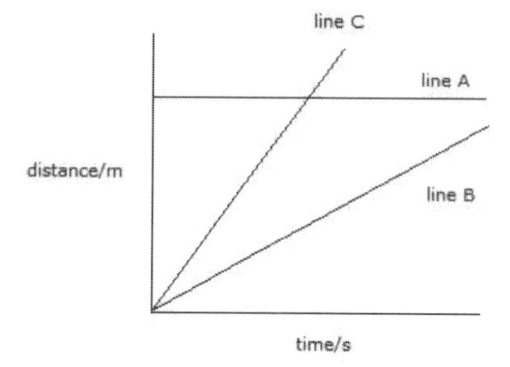

[Graph: distance/m vs time/s showing line A (horizontal), line B (moderate gradient from origin), line C (steep gradient from origin)]

If the gradient is zero, this means that the speed is zero, so line A represents an object that is stationary. From earlier, line B and C represent constant speeds with line C representing an object with a greater constant speed because it has a greater gradient (steeper line).

Figure 5 below shows another sketch of a distance-time graph representing the motion of another object. Initially the object is travelling at a constant speed as shown by the straight line. The line then gradually reduces in its steepness which shows that it is decelerating (slowing down). The horizontal line at the end then indicates that the object is stationary.

Figure 5

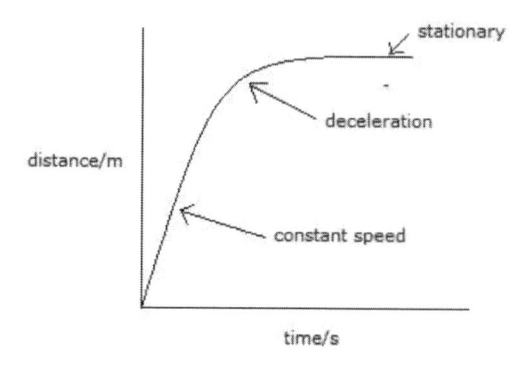

[Graph: distance/m vs time/s showing constant speed section, deceleration section, and stationary section]

If we now consider the velocity-time graph shown in figure 6:

FIGURE 6: A GRAPH OF VELOCITY AGAINST TIME

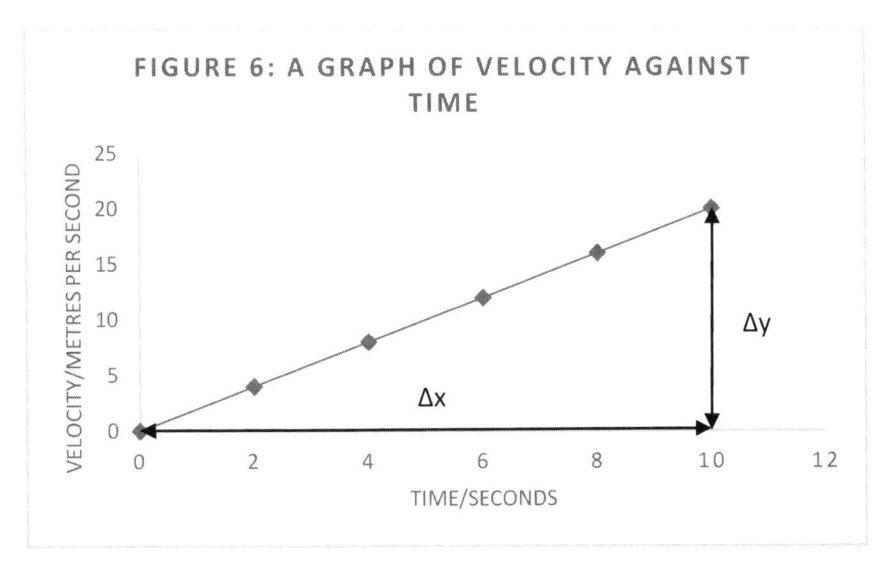

In a similar way to how we calculated the speed from a distance-time graph, the gradient of a velocity-time graph gives us the acceleration.

In the graph above, **Δy = 20m/s** and **Δx = 10s**. Therefore the gradient of this graph is:

$$gradient = \frac{\Delta y}{\Delta x}$$

$$gradient = \frac{20 m/s}{10 s}$$

gradient = 2m/s².

In this example, from the gradient formula, we have divided a velocity of 20m/s by a time of 10s. Looking at the units, we have divided **metres per second** by **seconds**. This is why the gradient of a velocity-time graph is equal to the acceleration. The units for the gradient are **m/s²** which are the same as the units for acceleration.

(**Note:**
This can be proved as follows:

$$\frac{m}{s} \div s$$

Is the same as,

$$\frac{m}{s} \div \frac{s}{1}$$

(we have made **s** into a fraction by writing s/1, just as we could make the number 5 a fraction by writing 5/1)
When dividing fractions we change the division sign into a multiplication sign and turn the second fraction upside down. This will give:

$$\frac{m}{s} \times \frac{1}{s}$$

When multiplying fractions we multiply the numerators together (m x 1 = m) and multiply the denominators together (s x s = s²). This then simplifies to **m/s²**)

Therefore, for a velocity-time graph:

The gradient of a velocity-time graph is equal to the acceleration.

Looking at the graph in figure 6 you can see that the gradient does not change, it is constant. If the gradient is the same, this means that the acceleration stays the same. **Therefore a graph like the one in figure 6 where the gradient does not change shows that the object is travelling at a constant acceleration.**

Also, if the gradient of a velocity-time graph is equal to the acceleration we can also conclude the following:

The greater the gradient (the steeper the line) the greater the acceleration.

If we now consider the velocity-time graph in figure 7:

The distance travelled can also be calculated from a velocity-time graph. **The distance travelled is equal to the area under the graph.** Looking at the graph in figure 7, this shows an object accelerating (speeding up) for the first 4 seconds, then travelling at a constant velocity for 4 seconds and then decelerating (slowing down) for 2 seconds at which point the velocity is zero (stationary). The distance travelled during this time can be calculated from figure 7 by calculating areas A, B and C and then adding the three areas together.

Area A is a triangle. The formula for a triangle is given below:

$$area\ of\ triangle = \frac{1}{2} \times base \times height$$

Or,

$$area\ of\ triangle = \frac{base \times height}{2}$$

Both of these are equivalent to each other and you can use whichever one you prefer.

FIGURE 7: A GRAPH OF VELOCITY AGAINST TIME

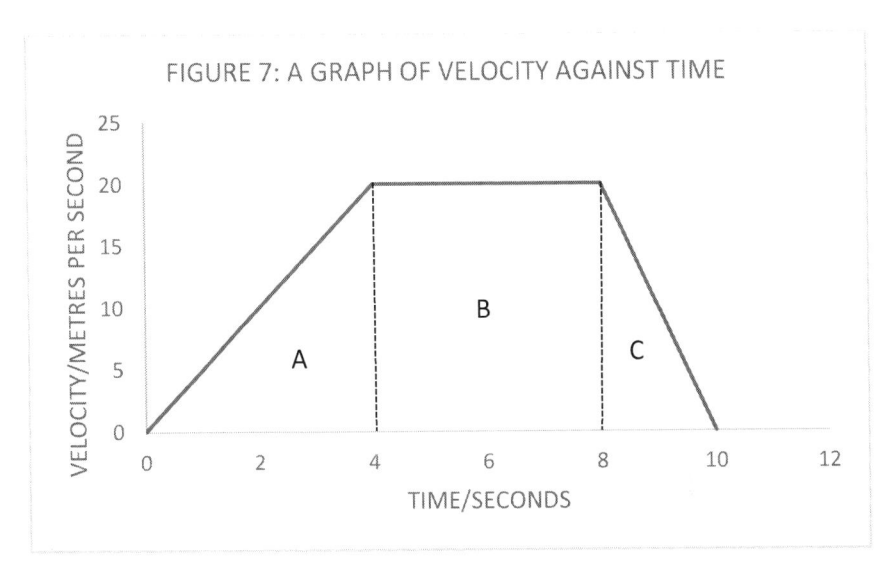

(Velocity/metres per second vs Time/seconds. The graph shows velocity rising linearly from 0 at t=0 to 20 at t=4 (region A), staying constant at 20 from t=4 to t=8 (region B), and decreasing linearly from 20 at t=8 to 0 at t=10 (region C).)

Area A

$$\text{area of triangle A} = \frac{\text{base} \times \text{height}}{2}$$

$$\text{area of triangle A} = \frac{4 \times 20}{2}$$

$$\text{area of triangle A} = \frac{80}{2}$$

Area of triangle A = 40m

Note:
The units are metres, m, because we have multiplied the base, which is time, by the height, which is velocity. If we multiply the units of time and velocity together we get:

$$s \times \frac{m}{s}$$

Here the seconds, **s**, cancel and we are left with **m**, which is metres. To show why these cancel let's consider some numbers. If we replace **s** with the number **5** and **m** with the number **10** then we have:

$$5 \times \frac{10}{5}$$

5 x 10 = 50 and 50 ÷ 5 = 10. We could also say that the 5's will cancel to give 10 or we could also say that 10 ÷ 5 = 2 and 5 x 2 = 10. Basically, we are just left with the number 10, which if we go back to the units, we just let m = 10, so the number 10 represents **m**.

The same rules apply to letters as well as numbers and this is why when multiplying seconds, **s** by **m/s** we are left with the units of **m**, metres.

Area B

Area B is a rectangle. The area of a rectangle is given below:

$$area\ of\ rectangle = length \times width$$

Therefore,
$$area\ of\ rectangle\ B = 20 \times 4$$

Area of rectangle B = 80m.

Area C

Area C is another triangle. The area of a triangle is:

$$area\ of\ triangle = \frac{base \times height}{2}$$

$$area\ of\ triangle\ C = \frac{2 \times 20}{2}$$

$$area\ of\ triangle\ C = \frac{40}{2}$$

Area of triangle C = 20m

Therefore, the total distance travelled by the object in figure 7 equals:

Distance travelled = 40 + 80 + 20 = 140m

If we take a look at the section from 8 to 10 seconds on the graph in figure 7 this shows a deceleration. If we calculate the gradient of this section we get:

$$gradient = \frac{\Delta y}{\Delta x}$$

$$gradient = \frac{-20}{2}$$

$$gradient = -10 m/s^2$$

Remembering that the gradient = acceleration, this means that the acceleration = $-10 m/s^2$. The negative sign shows that it is a deceleration!

Finally, if we consider the sketch of a velocity-time graph given in figure 8. From what we have discussed earlier, line A represents an object moving with a constant acceleration. Line B is also a constant acceleration but because the gradient is greater (the line is steeper) than the gradient of line A, object B has a greater acceleration than object A. **Line C represents a constant velocity!**

Figure 8

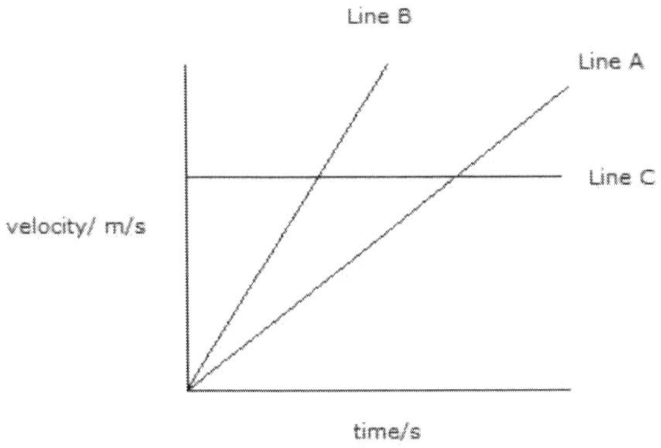
Line B
Line A
Line C
velocity/ m/s
time/s

If we now consider the current–voltage graph in figure 9:

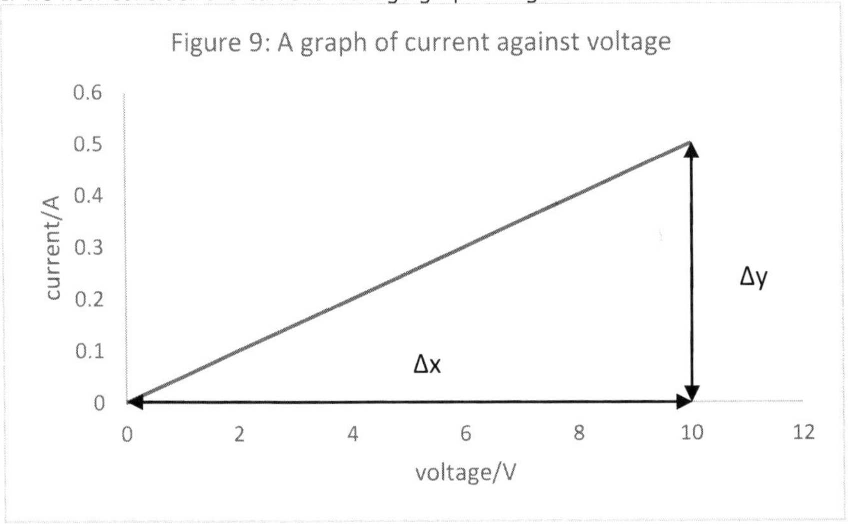
Figure 9: A graph of current against voltage
0.6
0.5
0.4
0.3
0.2
0.1
0
current/A
Δy
Δx
0
2
4
6
8
10
12
voltage/V

The gradient of this graph will be:

$$gradient = \frac{\Delta y}{\Delta x}$$

$$gradient = \frac{0.5A}{10V}$$

gradient = 0.05A/V

If we look at the units of the gradient, we have **A/V** (amps per volt). Looking back to section 30 however, $\boldsymbol{V = I\ x\ R}$. Rearranging to make **R** the subject of the formula will give:

$$R = \frac{V}{I}$$

If we consider the units on the right hand side of this formula we have a voltage divided by a current which gives **V/A** (volts per amp, the SI unit though is the ohm,Ω). Therefore, if you want to find the resistance from a current-voltage graph you do the following:

$$Resistance = \frac{1}{gradient}$$

The right hand side of this formula can be shown to give equivalent units of the ohm, Ω, as follows:

The gradient has units of **A/V**, therefore this gives:

This is the same as:

$$\frac{1}{\frac{A}{V}}$$

$$1 \div \frac{A}{V}$$

If we write the number 1 as a fraction this gives:

$$\frac{1}{1} \div \frac{A}{V}$$

When dividing fractions we change the division sign into a multiplication sign and turn the second fraction upside down. This will give:

$$\frac{1}{1} \times \frac{V}{A}$$

When multiplying fractions we multiply the numerators together (1 x V = V) and multiply the denominators together (1 x A = A). This then simplifies to **V/A** which is equivalent to the ohm, which is the unit of resistance. Therefore the resistance from the graph in figure 9 is:

$$Resistance = \frac{1}{gradient}$$

$$Resistance = \frac{1}{0.05}$$

Resistance = 20V/A = 20Ω

Looking at the formula for calculating the resistance from the gradient we have:

$$Resistance = \frac{1}{gradient}$$

This means that for a current-voltage graph, the greater the gradient, the smaller the resistance and the smaller the gradient the greater the resistance. If the gradient was 2A/V for example, the resistance would equal 1/2 which equals 0.5Ω. If the gradient was 0.5A/V, the resistance would equal

1/0.5 which equals 2Ω.

Note:
The graph in figure 9 is an example of an ohmic conductor because it obeys Ohm's law. Ohm's law can be stated as:
The current flowing through an ohmic conductor is directly proportional to the potential difference (voltage) across it, provided the temperature and other physical conditions remain constant.
It is important to note that for current to be directly proportional to potential difference (voltage) the graph must be a straight line through the origin (0,0).

If current and voltage are directly proportional, this means that, for example, if the current is doubled the voltage is doubled. If the current is halved, the voltage is halved. If the current is tripled, the voltage is tripled.

If we now consider the extension-force graph shown in figure 10:

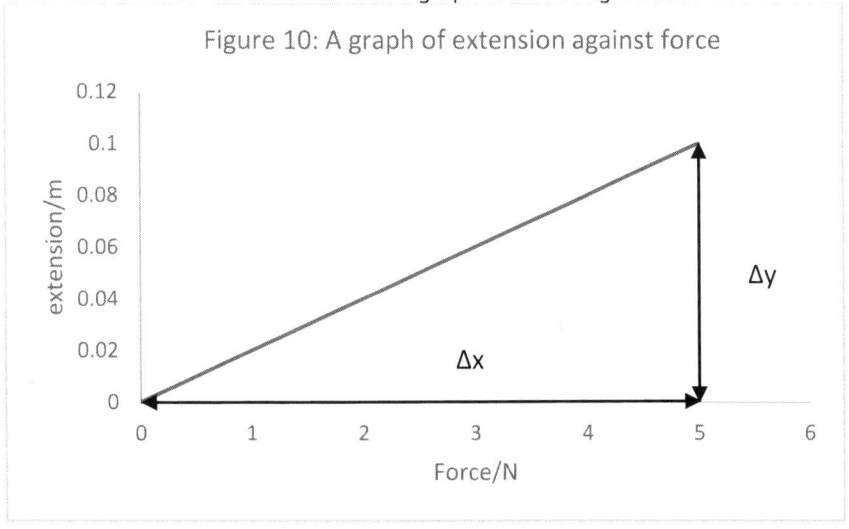

Figure 10: A graph of extension against force

From section 20, $F = k \times e$. Rearranging this to make **k** the subject of the formula will give:

$$k = \frac{F}{e}$$

The spring constant, **k** having units of N/m.
The gradient of this graph however will have units of m/N! This is the same situation that we had with the current-voltage graph! Therefore, using the same theory, the spring constant **k**, can be calculated from a graph like the one in figure 10 as follows:

$$spring\ constant, k = \frac{1}{gradient}$$

Therefore,

$$gradient = \frac{\Delta y}{\Delta x}$$

$$gradient = \frac{0.10}{5}$$

$$gradient = 0.02 m/N$$

But,
$$spring\ constant, k = \frac{1}{gradient}$$

Therefore,
$$spring\ constant, k = \frac{1}{0.02}$$

Spring constant, k = 50N/m.

Looking at the formula for calculating the spring constant from the gradient we have:
$$spring\ constant, k = \frac{1}{gradient}$$

This means that for an extension-force graph, the greater the gradient, the smaller the spring constant and the smaller the gradient the greater the spring constant. If the gradient was 0.04m/N for example, the spring constant would equal 1/0.04 which equals 25N/m. If the gradient was 0.02m/N, the spring constant would equal 1/0.02 which equals 50N/m.

Questions

1) Figure 11 below shows a distance-time graph for a cyclist. Calculate the speed of the cyclist.

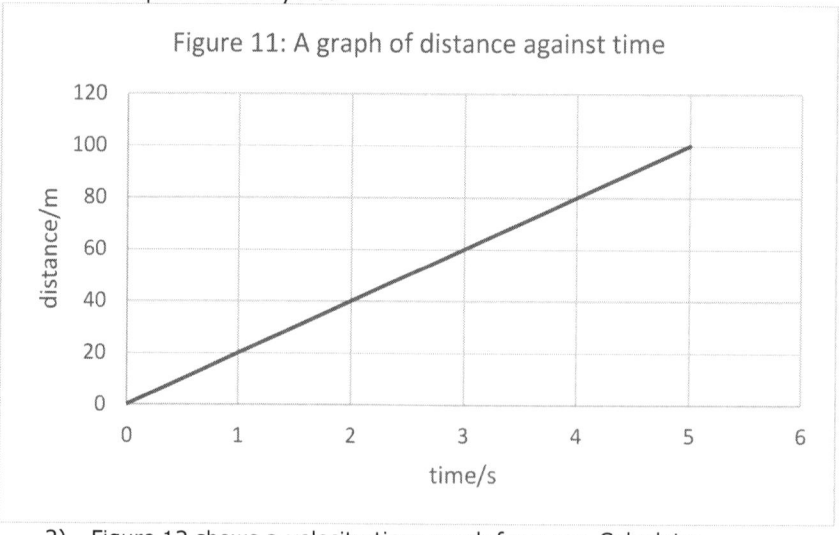

Figure 11: A graph of distance against time

2) Figure 12 shows a velocity-time graph for a car. Calculate:
   a) The acceleration during the first 6 seconds.
   b) The distance travelled after 12 seconds.
3) Figure 13 shows a velocity-time graph for a cyclist. Calculate:
   a) The acceleration during the first 5 seconds.
   b) The deceleration during the final 10 seconds.
   c) The total distance travelled.

Figure 12: A graph of velocity against time

Figure 13: A graph of velocity against time

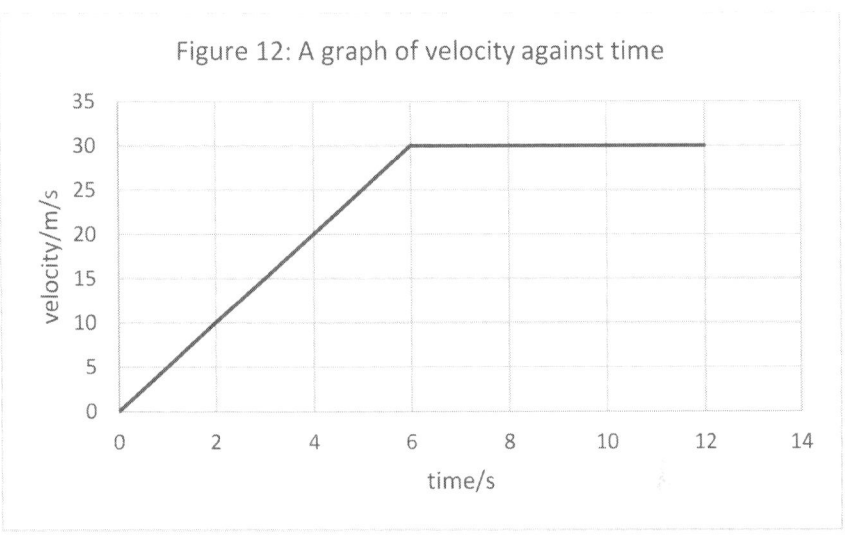

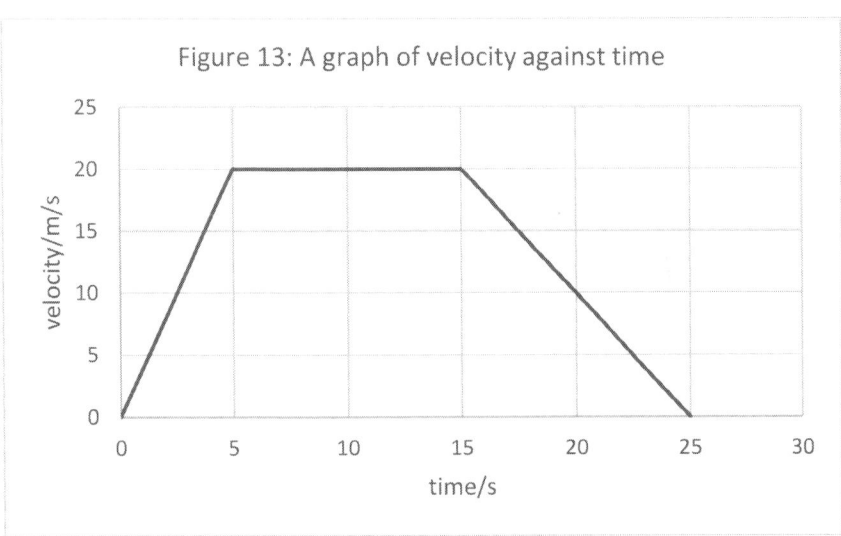

4) Using the table below, plot a graph of distance against time (the quantity before against is on the y-axis and the quantity after against is on the x-axis) and calculate the speed.

time/s	distance/m
0	0
2	5
4	10
6	15
8	20
10	25
12	30

5) Using the table below, plot a graph of velocity against time and calculate:
a) The acceleration during the first 3 seconds.
b) The distance travelled after 8 seconds.

time/s	velocity/m/s
0	0
1	6
2	12
3	18
4	18
5	18
6	18
7	18
8	18

6) Figure 14 below shows a current-voltage graph of a resistor. Calculate the resistance of the resistor.

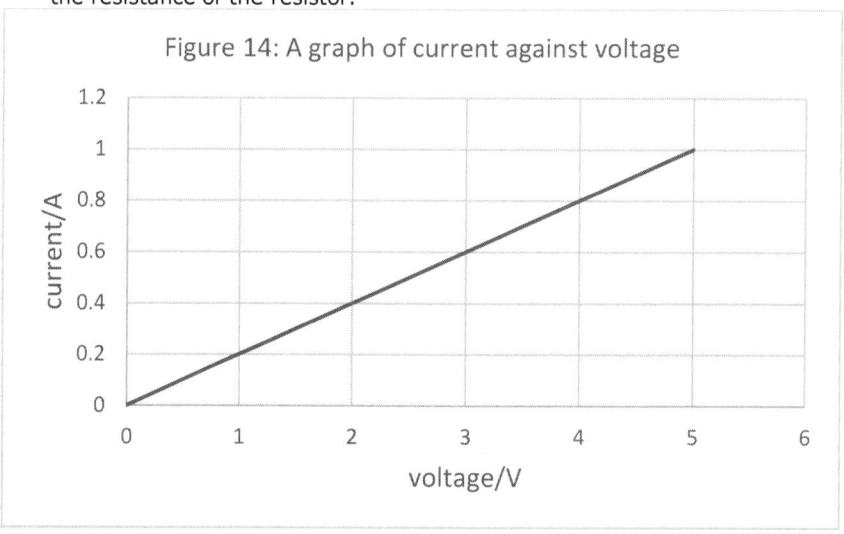

Figure 14: A graph of current against voltage

7) Figure 15 shows an extension-force graph of a spring. Calculate the spring constant of the spring.

Figure 15: A graph of extension against force

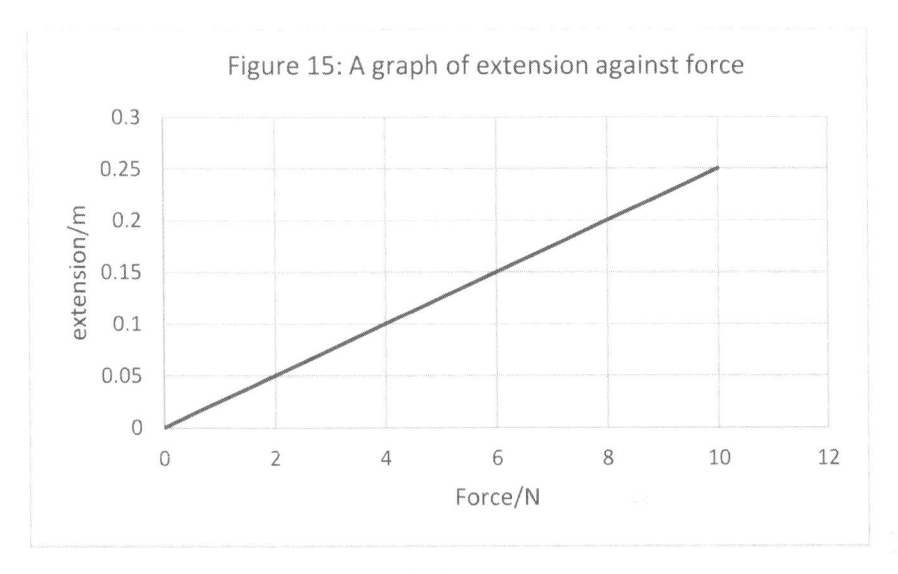

[Graph showing extension/m (y-axis, 0 to 0.3) vs Force/N (x-axis, 0 to 12), linear line from origin through approximately (10, 0.25)]

50. Velocity

The formula for velocity is given below:

$$velocity = \frac{displacement}{time}$$

Velocity is the speed in a given direction measured in metres per second, m/s.
Displacement is the distance in a given direction measured in metres, m.
Time is measured in seconds, s.

Both velocity and displacement are known as vectors because they have both magnitude (size) and direction. For example a velocity might be 10m/s North. The '10' is the magnitude (or size) of the vector and 'North' is the direction. Time is known as a scalar, as it only has magnitude, for example 20s. The '20' gives you the magnitude (or size) of the time.

Displacement can be illustrated in figure 1 below:

Figure 1

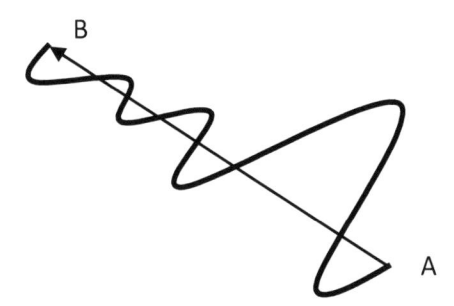

[Diagram showing a wiggly line from point A to point B with an arrow]

Imagine that the 'wiggly' line in figure 1 shows the route taken in a road race that was 10km long. The distance of the road race would be 10km. The

displacement however is the distance travelled in a given direction. It is the distance from 'start to finish' in a given direction. The displacement from A to B might be 5km North-West, for example, and is shown in figure 1 by the straight line from A to B. Therefore in the above example:
Distance = 10km
Displacement = 5km North-West.

Another important point to remember is that, for example, if a 400m runner runs a 400m race and stopped in the exact point at which they started, the distance that they had covered would be 400m, but the displacement would be 0m. They have started and finished at the same point, so the displacement would be 0m!

The formula for velocity can be written as:

$$v = \frac{s}{t}$$

where **v** is the velocity, **s** is equal to the displacement and **t** is the time.

(**Note:**
When using this formula it is important to remember that **s** is the displacement not the speed! If you have to use both the speed formula and the velocity formula, instead of writing the speed formula as:

$$s = \frac{d}{t}$$

It may be worth writing it as:

$$speed = \frac{d}{t}$$

in order that you don't get 'speed' and 'displacement' mixed up).

Taking a look at rearranging the formula, if we want to make **s** the subject of the formula we need to multiply both sides by **t**. This will give:

$$v \times t = s$$

Therefore,

$$s = v \times t$$

From $s = v \times t$, if we want to make **t** the subject of the formula we need to divide both sides by **v**. This gives:

$$\frac{s}{v} = t$$

Therefore,

$$t = \frac{s}{v}$$

Example 1

A 400m runner runs 400m in 50s. If the runner starts and finishes at the same point, calculate:
    a) The speed

b) The velocity

a) speed = ?
   d = 400m
   t = 50s

$$speed = \frac{d}{t}$$

$$speed = \frac{400}{50}$$

speed = 8m/s

b) v = ?
   t = 50s
   s = 0m (the runner has started and finished at the same point, so the displacement is 0m.

$$v = \frac{s}{t}$$

$$v = \frac{0}{50}$$

v = 0m/s

Example 2

A car travels at a velocity of 40m/s North for 20s. Calculate the displacement.

v = 40m/s North
t = 20s
s = ?

$$v = \frac{s}{t}$$

$$s = v \times t$$

s = 40 x 20

s = 800m North

It is important to remember that displacement is a vector and therefore must have magnitude and direction.

Example 3

A cyclist has a displacement of 600m South. If the cyclist had a velocity of 20m/s South during this displacement, how long did it take for this displacement?

s = 600m South
v = 20m/s South
t = ?

$$v = \frac{s}{t}$$

$$t = \frac{s}{v}$$

$$t = \frac{600}{20}$$

$$t = 30s$$

Questions

1) A sprinter runner runs 100m West in 9.80s. Calculate the velocity.
2) A cyclist travels 500m North-East in 20s. Calculate the velocity.
3) An 800m runner completes an 800m race in 1 minute 45 seconds. If he starts and finishes at the same point, calculate:
   a) His speed
   b) His velocity
4) A runner has a velocity of 7m/s South-East. Calculate the displacement of the runner if the time at this velocity is 15s.
5) A Go-kart has a velocity of 15m/s East. If the displacement is 375m East, calculate the time taken for this displacement.
6) A car travels at a velocity of 30m/s North for 1 minute. Calculate the displacement.
7) A car has a displacement of 100km North. Calculate the velocity if the time for this displacement is 50.0 minutes.

51. Answers to questions

If an answer has another answer after it in brackets with a * (i.e. 4167m *) this is the answer where answers needed during the calculation have not been rounded or it is the unrounded answer.

1. Powers of 10
1a) 40,000,000 1b) 30,000 1c) 5,370,000 1d) 0.000045 1e) 0.000000000067 1f) 25,000 1g) 0.0000051 1h) 714,000,000 1i) 0.00904 1j) 200,000

2. Prefixes
1) 75,000N 2) 0.000050s 3) 5,000,000V 4) 0.004A 5) 10,000,000,000J 6) 0.008m 7) 0.000000006C 8) 0.60m

5. Changing the subject of a formula
a) $a = F/m$ b) $t = Q/I$ c) $g = W/m$ d) $t = d/s$ e) $d = s \times t$ f) $s = v^2 - u^2/2a$ g) $t = 2s/(u + v)$

7. Speed
1) 8.77m/s 2) 50km/h 3) 1.67m/s 4) 1000m 5) 2160m 6) 1000m 7) 13.3s 8) 130s 9) 600m 10) 0.002m/s (2 x 10^{-3}m/s) 11) 4160m (4167m *) 12) 2 hours 13) 500s 14) 27 x 10^3s (2 s.f.) (27,000s) 15) 28.16 x 10^6m

8. Acceleration
1) 12m/s 2) 4m/s² 3) 8s 4) 10m/s 5) 6s 6) 45m/s 7) 7.44m/s² (7.45m/s² *) 8) 2.5m/s² 9) 3s 10) 17.1m/s²

9. Weight
1a) 20N 1b) 50N 1c) 100N 1d) 50,000N 1e) 300,000N 2a) 20kg 2b) 50kg 2c) 1000kg 2d) 4000kg 2e) 20,000kg 3a) 4N 3b) 0.64N

4) 2.8 x 10⁶kg (2,800,000kg)   5) 1978kg

10. F = ma
1) 12N   2) 6m/s²   3) 3750N   4) 80N   5) 7000N   6) 5m/s²   7) 2.32m/s²
8) 7.5 x 10³kg (7500kg)   9a) 2000N   9b) 5000N   10a) 5m/s²   10b) 25m/s

11. Momentum
1a) 6kgm/s East   1b) 12kgm/s South   1c) 30kgm/s North
1d) 0.12kgm/s West   1e) 0.15kgm/s North   2a) 5m/s South   2b) 10m/s East
2c) 4m/s West   3) 22.2 x 10⁴kgm/s East (222,000kgm/s East)
4) 5 x 10⁻³kgm/s North (0.005kgm/s North)   5) 75kg

12. Force = rate of change of momentum
1a) 20N   1b) 50N   1c) 400N   1d) 4000N   1e) 25N   2a) 2000Ns
2b) 1200Ns   2c) 100,000Ns   2d) 5Ns   2e) 125Ns   3) 80N   4a) 10,000N
4b) 10m/s²   5) 6s

13. Conservation of momentum
1) 800m/s   2) 4m/s   3) 0.67m/s   4) 21.7m/s West   5) 6m/s

14. Work done
1) 840J   2) 125J   3) 700J   4) 60J   5) 10m   6) 0.50kg   7) 0.900m
8) 3825J

15. Kinetic energy
1) 100J   2) 20.3J (20.25J *)   3) 490J   4) 2000kg   5) 3m/s   6) 20m/s
7) 4.4 x 10⁹J (4.3904 x 10⁹J *)   8) 6.10 x 10⁸J (6.11 x 10⁸J *)   9) 17m/s
10) 9.5 x 10³kg (9500kg)

16. Gravitational potential energy
1a) 500J   1b) 1250J   1c) 5.0 x 10⁴J (50,000J)   1d) 1.0 x 10⁵J (100,000J)
1e) 1.5 x 10⁵J (150,000J)   2a) 5m   2b) 15m   2c) 35m   2d) 40m
2e) 2.5km (2500m)   3) 30m   4a) 800N   4b) 80kg
5) 1.4 x 10¹⁰J (1.428 x 10¹⁰J *)   6) 30m   7) 15J   8) 40m

17. Conservation of energy
1a) 120J   1b) 120J   1c) 8.9m/s   1d) 90J   1e) 90J   2) 50J   3a) 500J
3b) 500J   3c) 14.1m/s   3d) 400J   3e) 12.6m/s   4a) 25J   4b) 1.25m
5a) 400,000J   5b) 400,000J   5c) 89.4m/s

18. Moments (or Torques)
1a) 0.75Nm   1b) 1.5Nm   1c) 1.8Nm   1d) 3.0Nm   1e) 4.5Nm   2a) 2.00m
b) 1.50m   2c) 1.00m   2d) 0.750m   2e) 0.500m   3) 20N   4a) 10m   4b) 5m
4c) 0.50m   4d) 1.5m   4e) 3.0m   5) 2.4Nm

19. Law of moments
1) 400N   2) 2.0m   3) 3.0m   4) 3m   5)1600N

20. F = k x e
1) 0.72N   2) 50N/m   3a) 0.20m   3b) 0.30m   3c) 0.40m   3d) 0.50m
3e) 0.25m   4a) 40N/m   4b) 0.10m   5a) 50N/m   5b) 0.50m   6a) 4.0N
6b) 8.0N   6c) 12N   6d) 20N   6e) 32N

21. Centripetal Force
1a) 200N   1b) 50N   1c) 100N   1d) 25N   1e) 40N   2a) 0.16kg   2b) 0.47kg
2c) 1.6kg   2d) 2.3kg   2e) 3.1kg   3a) 0.20m   3b) 0.45m   3c) 0.80m
3d) 1.3m (1.25m *)   3e) 5m   4) 25m/s   5) 80kg
6) 1.0 x 10³m/s (1022m/s *)   7) 3.6 x 10⁷m (36,000,000m)

22. Density
1a) 0.136kg   1b) 5420kg   1c) 27,100kg   1d) 0.0542kg
1e) 2.71 x 10⁻³kg (0.00271kg)   2a) 2.24 x 10⁻⁶m³   2b) 5.60 x 10⁻³m³
2c) 0.11m³   2d) 6.7 x 10⁻⁶m³   2e) 4.5 x 10⁻⁴m³   3) 3300kg/m³
4) 4.4 x 10⁻⁴m³   5) 10,500kg/m³   6) 0.6773kg   7) 0.6998m³

23. Pressure
1a) 50,000Pa   1b) 20,000Pa   1c) 10,000Pa   1d) 6.67 x 10³Pa   1e) 5000Pa
2a) 40N   2b) 800N   2c) 1000N   2d) 60N   2e) 2000N   3a) 0.020m²

3b) 0.13m² 3c) 0.050m² 3d) 0.40m² 3e) 1.0m² 4) 0.25m²
5) 2.5 x 10⁶Pa (2,500,000Pa) 6) 7.50 x 10⁵Pa (750,000Pa) 7) 300N

24. Pressure variation with depth in fluids
1a) 50,000Pa 1b) 200,000Pa 1c) 50,000Pa 1d) 4000Pa 1e) 5000Pa
2a) 2.0m 2b) 0.50m 2c) 0.25m 2d) 0.20m 2e) 0.10m 3) 2000m
4) 13.5 x 10³kg/m³ (13,546kg/m³ *) 5) 8.7 x 10²Pa (870Pa)

25. Hydraulics
1a) 100N 1b) 200N 1c) 250N 1d) 500N 1e) 2.5 x 10³N (2500N)
2a) 400N 2b) 800N 2c) 600N 2d) 1.0 x 10³N (1000N)
2e) 1.2 x 10³N (1200N) 3) 400cm² 4a) 125N/cm² 4b) 125N/cm²
4c) 2000N 5) 150N 6) 600cm² 7a) 200N/cm² 7b) 200N/cm²
7c) 3.6 x 10³N (3600N)

26. Boyle's law
1a) 1.5m³ 1b) 2.0m³ 1c) 1.0m³ 1d) 1.2m³ 1e) 0.75m³ 2) 857mmHg
3) 300cm³ 4) 250cm³ 5) 1.33 x 10⁵Pa

27. Pressure law
1a) 2.0 atmospheres 1b) 2.5 atmospheres 1c) 3.0 atmospheres
1d) 5.0 atmospheres 1e) 10 atmospheres 2a) 330K 2b) 375K 2c) 450K
2d) 420K 2e) 480K 3) 300K (27°C) 4) 1.5 x 10⁵Pa 5) 250K (-23°C)

28. Specific heat capacity
1a) 15.4 x 10³J (15,400J) 1b) 23.1 x 10³J (23,100J)
1c) 77.0 x 10³J (77,000J) 1d) 11.6 x 10⁴J (115,500J *) 1e) 15.4 x 10⁴J
(154,000J) 2) 50°C 3) 132J/kg/°C 4) 0.20kg 5) 30.6 x 10⁴J (306,000J)
6) 100°C 7) 3.4 x 10³J/kg/°C (3350J/kg/°C *)

29. Latent heat
1) 1.34 x 10⁶J (1.336 x 10⁶J *) 2) 4.52 x 10⁶J (4,520,000J) 3) 0.400kg
4) 1.67 x 10⁵J (167,000J) 5) 3.80 x 10⁵J/kg (380,000J/kg)
6) 4.52 x 10⁶J (4,520,000J) 7) 2.41 x 10⁶J 8) 1.5 x 10⁵J 9) 700J
10) 4.52 x 10⁵J (452,100J *)

30. V = I x R
1) 40V 2) 5A 3) 2 x 10³Ω (2000Ω) 4) 6V 5) 1.25 x 10⁻⁵A 6) 25Ω 7) 6V
8) 5.0 x 10⁻³A (0.0050A) 9) 3.5 x 10³Ω (3500Ω)
10) 8.0 x 10⁶Ω (8,000,000Ω)

31. Series circuits
1a) 5Ω 1b) 4A 1c) $V_1 = 8V, V_2 = 12V$ 2a) 10Ω 2b) 3A 2c) $V_1 = 12V$,
$V_2 = 18V$ 3a) 5kΩ (5000Ω) 3b) 3 x 10⁻³A (0.003A) 3c) $V_1 = 6V, V_2 = 9V$
4a) 6Ω 4b) 1A 4c) $V_1 = 1V, V_2 = 2V, V_3 = 3V$ 5a) 12Ω 5b) 2A
5c) $V_1 = 4V, V_2 = 4V, V_3 = 6V, V_4 = 10V$

32. Parallel circuits
1a) 2V 1b) $I_1 = 0.5A, I_2 = 4A$ 1c) 4.5A 2a) 10V 2b) $I_1 = 5A, I_2 = 5A$
2c) 10A 3a) 8V 3b) $I_1 = 8 \times 10^{-3}A$ (0.008A), $I_2 = 4 \times 10^{-3}A$ (0.004A)
3c) 0.012A 4a) 30V 4b) $I_1 = 0.03A, I_2 = 6 \times 10^{-3}A$ (0.006A),
$I_3 = 3 \times 10^{-3}A$ (0.003A) 4c) 0.039A 5a) 10V 5b) $I_1 = 0.01A$,
$I_2 = 2 \times 10^{-3}A$ (0.002A), $I_3 = 1 \times 10^{-3}A$ (0.001A) 5c) 0.013A 6a) 6V 6b) 3V

33. Q = I x t
1a) 3C 1b) 30C 1c) 180C 1d) 10,800C 1e) 54,000C 2a) 0.054A
2b) 0.027A 2c) 9 x 10⁻³A (0.009A) 2d) 4.5 x 10⁻³A (0.0045A) 2e) 0.018A
3) 3600s 4) 36,000s 5) 14,400C 6) 4 x 10⁻³A (0.004A) 7) 86,400s

34. E = V x Q
1a) 40J 1b) 80J 1c) 160J 1d) 400J 1e) 72J 2a) 10V 2b) 20V 2c) 5V
2d) 2V 2e) 4V 3a) 2C 3b) 10C 3c) 30C 3d) 70C 3e) 60C 4) 15J
5) 9V 6) 40C 7) 6V

35. P = V x I
1) 12W 2) 30A (29.6A *) 3) 1000W (1000.5W *) 4) 3.48A
5) 150W (149.5W *) 6) 3.91A 7) 9.57A 8) 2.5W 9) 1W 10) 0.26A

36. E = P x t
1) 1.2×10^3W (1200W) 2a) 6000J 2b) 1.8×10^5J (180,000J)
2c) 3.6×10^5J (360,000J) 2d) 7.2×10^5J (720,000J)
2e) 3.6×10^6J (3,600,000J) 3) 5400s 4) 60W 5) 54,000J 6) 28,800s
7) 3.6×10^6J (3,600,000J)

37. Paying for electricity
1a) 0.04kWh 1b) 0.2kWh 1c) 0.4kWh 2) 4p 3) 5p 4) 0.32p 5) 4p
6) 288p 7) 1.7p (1.67p *) 8) 10p 9) 18p (17.5p *) 10) 1p

38. E = V x I x t
1) 240J 2) 9000s 3) 29.6A 4) 6V 5) 9.60×10^4J (96,048J *) 6) 180s
7) 5.2A

39. Efficiency
1a) 133J (133.3J *) 1b) 667J (666.7J *) 1c) 2400J 1d) 4000J
1e) 5.33×10^6J 2a) 40J 2b) 200J 2c) 300J 2d) 24,000J 2e) 2800J
3) 0.75 (or 75%) 4) 1.75×10^3W (1750W) 5) 3.5×10^6J
6) 1.0×10^3W (1000W) 7) 0.30 (or 30%)

40. Transformers
1) 100V 2) 575 turns 3) 4015A 4) 30 turns 5) 320 turns 6) 2.5V 7) 2A

41. v = f x λ
1a) 10m/s 1b) 40m/s 1c) 100m/s 1d) 5m/s 1e) 2.5m/s 2a) 5Hz
2b) 4Hz 2c) 2Hz 2d) 40Hz 2e) 10Hz 3a) 0.05m 3b) 0.1m 3c) 0.02m
3d) 5.0×10^{-3}m (0.0050m) 3e) 1.0×10^{-3}m (0.0010m) 4) 7.5×10^{14}Hz
5) 433m (432.9m *) 6) 1×10^{10}Hz 7) 0.15m/s

42. T = 1/f
1) 0.10s 2) 10Hz 3) 2Hz 4) 2×10^{-3}s (0.002s) 5) 50Hz
6) 1×10^{-3}s (0.001s) 7) 1×10^{-9}s 8) 1×10^{-23}s 9) 7.52×10^{14}Hz
10) 5×10^{-7}s

43. Refractive index
1) 1.33 2) 74.6° 3) 25.5° 4) 1.31 5) 1.3×10^8m/s (1.25×10^8m/s *)
6) 1.28 7) 39.3° 8) 45.0°

44. Total internal reflection and critical angle
1) 49.8° 2) 1.28 3) 34.6° 4) 1.54 5) 1.33 6) 47.3° 7) 1.44

45. P = 1/f
1a) 1.0D 1b) 2.0D 1c) 4.0D 1d) 100D 1e) 1.25D 2a) -20D 2b) -2D
2c) -5D 2d) -1.67D 2e) -66.7D 3a) 0.125m 3b) -0.25m 3c) 0.067m
3d) 0.0100m 3e) -0.020m 4) 143D (142.9D *) 5) -3.3D 6) -0.025m
7) 0.022m

46. 1/u + 1/v = 1/f
1) 24cm 2) 15cm 3) -16.7cm 4) 12mm 5) 24cm 6) 60mm 7) 60cm

47. Magnification
1) 8cm 2) 16mm 3) 0.50 4) 0.25 5) 20mm 6) 20cm 7) 0.1

48. Half-life
1) 125c/s 2) 1/16 3) 1/32 4) 75c/s 5) 25c/s 6) 1/64
7b) 2.0 minutes (1.9 – 2.1 minutes)

49. Graphs
1) 20m/s 2a) 5m/s^2 2b) 270m 3a) 4m/s^2
3b) 2m/s^2 (or acceleration = -2m/s^2) 3c) 350m 4) 2.5m/s 5a) 6m/s^2
5b) 117m 6) 5Ω 7) 40N/m

50. Velocity
1) 10.2m/s West 2) 25m/s North-East 3a) 7.62m/s 3b) 0m/s
4) 105m South-East 5) 25s 6) 1.8×10^3m North (1800m North)
7) 33.3m/s North

About the Author

Brian Mills, CPhys, MInstP, is a chartered physicist, member of the Institute of Physics, and is currently head of physics at the highly rated John Taylor High School in Staffordshire, England. Prior to starting at John Taylor, Mills taught for two years at the Pingle School in Derbyshire.

In 1997 he graduated with first class honors in physics and sports science from Loughborough University and completed his postgraduate certificate in education (PGCE) the following year.

Mills is a former professional soccer player with Port Vale Football Club. He represented England at the U19 level in the 1991 World Youth Championships in Portugal before a spinal cord injury ended his career.

Today his hobbies include weight training, swimming, and astronomy. He lives near Lichfield, in Staffordshire, with his wife and two children.

Published by Physics Education Limited

Brian Mills

Physics calculations for GCSE & IGCSE

Brian Mills

24036876R00111

Printed in Great Britain
by Amazon